# Maple V

## Learning Guide

**Springer**
*New York*
*Berlin*
*Heidelberg*
*Barcelona*
*Budapest*
*Hong Kong*
*London*
*Milan*
*Paris*
*Santa Clara*
*Singapore*
*Tokyo*

# Maple V

## Learning Guide

K. M. Heal   M. L. Hansen   K. M. Rickard

With the Editorial Assistance of J. S. Devitt
Based in Part on the Work of B. W. Char

with 8 color figures

Waterloo Maple Inc.
450 Phillip St.
Waterloo, ON N2L 5J2
Canada

Camera-ready copy prepared using Springer-Verlag's TEX macros.
Printed and bound by Hamilton Printing Company, Rensselaer, NY.
Printed in the United States of America.

9 8 7 6 5 4 3 2 1

ISBN 0-387-98397-X Springer-Verlag New York Berlin Heidelberg SPIN 10659990
ISBN 0-387-98399-6 Springer-Verlag New York Berlin Heidelberg (Maple V software
boxed version) SPIN 10660056

# Contents

# Interactive Use of Maple

Maple V is a *Symbolic Computation System* or *Computer Algebra System*. Both phrases refer to Maple V's ability to manipulate information in a symbolic or algebraic manner. Conventional mathematical programs require numerical values for all variables. By contrast, Maple V maintains and manipulates the underlying symbols and expressions.

You can use these symbolic capabilities to obtain exact analytical solutions to many mathematical problems, including integrals, systems of equations, differential equations, and problems in linear algebra. Complementing the symbolic operations are a large set of graphics routines for visualizing complicated mathematical information, numerical algorithms for providing estimates and solving problems where exact solutions do not exist, and a complete and comprehensive programming language for developing custom functions and applications.

Maple V's extensive mathematical functionality is most easily accessed through its advanced worksheet-based graphical interface. A worksheet is a flexible document for exploring mathematical ideas and for creating sophisticated technical reports. Users of Maple have found myriad ways to exploit Maple's language and its worksheets.

Engineers and professionals in industries as diverse as agriculture and aerospace use Maple V as a productivity tool, replacing many traditional resources such as reference books, calculators, spreadsheets, and programming languages such as FORTRAN. These users gain quick answers to a wide range of day-to-day mathematical problems, creating projections and consolidating their computations into professional technical reports.

Researchers in many fields find Maple V to be an essential tool for their work. Maple is ideal for formulating, solving, and exploring mathematical models. Its symbolic manipulation facilities greatly extend the range of problems you can tackle.

Instructors use it to present lectures. Educators in high schools, colleges, and universities have revitalized traditional curricula by introducing problems and exercises that exploit Maple V's interactive mathematics. Students can concentrate on important concepts, rather than tedious algebraic manipulations.

The way in which you use Maple is in some aspects personal and dependent on your needs, but two modes are particularly prevalent.

The first mode is as an interactive problem-solving environment. When you work on a problem in a traditional manner, attempting a particular method of solution may take hours, and many pages of paper. Maple allows you to tackle much larger problems and frees you from mechanical errors. The interface provides documentation of the steps involved in finding your result. It allows you to easily modify a step or insert a new one in your solution method. Maple can then compute the new result effortlessly. Whether you are developing a new mathematical model or analyzing a financial strategy, you can learn a great deal about the problem you are tackling in very little time and with very little effort.

The second mode in which you may use Maple is as a system for generating technical documents. You can create interactive structured documents which contain live mathematics where you can change an equation and update the solution automatically. Maple's natural mathematical language allows easy entry of equations. You also can compute and display plots at will. In addition, you can structure your documents using modern tools such as styles, outlining, and hyperlinks, creating documents that are not only clear and easy to use, but easy to maintain. Since components of worksheets are directly associated with the structure of the document, you can easily translate your work into other typesetting languages, such as LaTeX or HTML.

Many types of documents can benefit from the features of Maple's worksheets. These facilities save you a lot of effort if you are writing a report or a mathematical book, and they are also appropriate for creating and displaying presentations and lectures. For example, outlining allows you to collapse sections to hide regions which contain distracting detail. Styles identify keywords and headings. Hyperlinks allow you to create live references which instantly transport the reader to pages containing related information. Above all, the interactive nature of Maple allows you to compute results and answer questions during presentations. You can clearly and effectively demonstrate why a seemingly acceptable solution method is inappropriate, or why a particular modification to a manufacturing process would lead to loss or profit.

This book is your introduction to Maple V. It systematically discusses important concepts and builds a framework of knowledge which will guide you in your use of the interface and the Maple language. This book introduces the most important commands and teaches you how to best use the on-line help system. More importantly, it presents the philosophy and methods of use intended by the designers of the system. These simple concepts allow you to use Maple fully and efficiently.

Whereas this book is a guide, the on-line help system is your reference manual. Maple's help system is handier than any traditional text as you can search for information in many ways, and the help facility is always at your fingertips.

This first chapter provides the essential information you need to begin using Maple V to solve problems and to effectively use the document-structuring features. Emphasis is on introducing the many aspects of the graphical interface. Going carefully through the four tutorials in this chapter will help you learn to interactively set up and solve mathematical problems, to enhance your worksheet to create a polished technical document, and to use the on-line help system. The remainder of this book discusses other principal areas of knowledge, beginning in chapter 2 with a more formal introduction to the Maple language.

## 1.1 The Worksheet Interface

The Maple V graphical interface includes most of the facilities that you expect in modern application software. For example, it supports standard mouse operations, including cutting and pasting. If you are comfortable with conventional software, such as word processors, you already have most of the knowledge you need to navigate Maple V's interface. You can perform standard operations such as opening, saving, and printing files in the manner with which you are already familiar.

On most platforms, you start Maple by double-clicking on the Maple icon. On others, you type the command *xmaple* or *maple* at a prompt. Maple starts and presents you with a new worksheet. Your Maple window will look like that shown in figure 1.1.

At the top of the window is the *menu bar* containing such menus as File and Edit. Immediately below the menu bar is the *tool bar*, which contains button-based shortcuts to common operations such as opening, saving, and printing. Immediately below the tool bar is the *context bar* which contains controls specific to the task you are currently performing. The next area is large and displays your worksheet, the region in which you

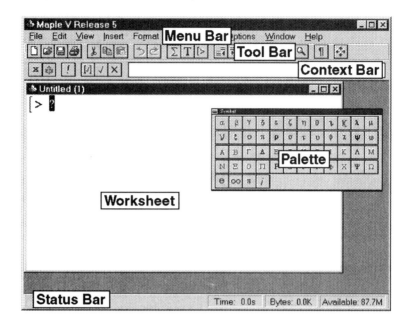

**FIGURE 1.1**  A NEW MAPLE SESSION

work. At the very bottom of the window is the *status bar*, which displays system information.

The *worksheet* is an integrated environment in which you interactively solve problems and document your work. Worksheets contain not only text and conventional document-processing information, but "live" mathematical commands and their automatically generated results.

At the prompt (> ) in the upper left corner there is a highlighted question mark (placeholder). (If not, set the option Input Display to Standard Math in the Option menu and then hit enter once.)

In addition to Maple commands and their results, you can include many other types of information in your document:

- You can include paragraphs of text. You have full control over the appearance of your paragraphs, even character by character, through the use of styles.

- You can include mathematical expressions and Maple commands within a paragraph.

- On some platforms, you can include *spreadsheets*, tables of mathematical values and/or symbolic formulae, which can be executed by Maple.

- You can include *hyperlinks*, special text regions that respond by jumping to another location in any worksheet or help page when clicked with the mouse.

- You can structure your document using styles, hyperlinks, and collapsible sections and subsections.
- On some platforms, you can embed live objects, such as figures and tables, from other application programs using OLE 2, the object linking and embedding standard under the Microsoft Windows family of operating systems.

However, simple problem-solving requires almost no special knowledge of the interface, and even sophisticated document presentation requires understanding only a few concepts.

## 1.2 Tutorial 1: Solving Problems

This tutorial takes you through a typical problem-solving session. It illustrates the following key interface concepts:

- entering and executing Maple commands
- editing existing Maple commands
- working with plots

As you start, please keep in mind to type in the expressions exactly as they appear on these pages. Since Maple is case-sensitive and typically uses lower case characters, take care when entering capital letters.

This tutorial guides you through some typical steps to compute the integral (anti-derivative)

$$\int x^2 \sin(x) dx$$

and explore the result.

1. Start Maple. On most platforms you can do this by double-clicking on the Maple icon. On others, you do this by typing the command *xmaple* or *maple* at a prompt. If Maple is already running, click on the ☐ button to generate a new worksheet.

2. Look for the Maple prompt, (> ), located near the top left of the large blank worksheet region. There will be a highlighted question mark (placeholder) in the input field, indicating that input is to be entered in *standard math* mode.

   Use the expression palette to create the integrand $x^2 \sin(x)$, by clicking on the ▣ template. Now you have two placeholders and the first is highlighted. Click on the ▣ template and that first placeholder gets replaced by two others. Type x and hit *tab*. Type 2 and hit *tab*. Click on

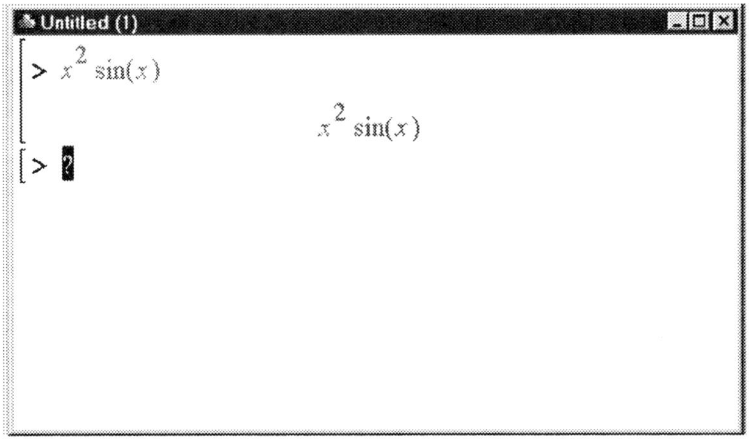

**FIGURE 1.2**   EXECUTION GROUP BRACKETS

**sin** to replace the last placeholder with a sine function, and type x and *enter* to complete the integrand. Hitting *enter* one more time creates the corresponding output field.

Notice the large square bracket just to the left of the original input prompt on your screen. This bracket groups the Maple input with its corresponding output. See figure 1.2.

3. Using your right-most mouse button (or holding down the *option* key if you are using a Mac), click directly on the output expression just created. A pop-up menu (called a *context menu*) with several choices appears. Choose `Integrate` from the menu and x from the submenu that appears. An automatically generated `int` command is inserted in the worksheet. (The R0 name is automatically generated by Maple.)

4. The palettes help you to create Maple input without knowing much about Maple's syntax. When the palette is used to create an expression, the *edit field* of the *context bar* is automatically updated with the associated command-line syntax. Observing the *edit field* can teach you about Maple notation, so that you can create more complicated commands in the future. The current command could be performed by switching to Maple notation mode and typing `int(x^2 * sin(x), x);`[1]

---

[1]In Maple Notation mode the command-line syntax requires a semi-colon or a colon as a *terminator* to complete the command.

Even though you could continue in standard math mode, enter the following two commands in command-line syntax. First, right-click (or use *option*-click on a Mac[2]) on the new input field and toggle off the Standard Math option. Notice that there is now no placeholder after the prompt, no *edit field* in the *context bar*, and that when you type, the characters appear directly in the input region.

5. Tell Maple to assign the value of the computed integral to a more representative name, answer, by entering the following command. The percentage character (%), known as the *ditto operator*, tells Maple to apply the new command to the previous successfully executed result.

> answer := % ;

$$answer := -x^2 \cos(x) + 2 \cos(x) + 2 x \sin(x)$$

Now you will be able to use the value of answer in subsequent commands.

6. The answer is an expression in terms of the unknown variable $x$, but you can tell Maple to evaluate answer at a particular $x$. (Remember to switch the input field to command-line mode first.)

> eval( answer, x=Pi/3 );

$$-\frac{1}{18} \pi^2 + 1 + \frac{1}{3} \pi \sqrt{3}$$

The command eval takes two parameters, the expression to evaluate and the point at which to evaluate the expression, respectively.

7. As you can see, Maple retains answers in symbolic form as long as it can; if you want to view an answer in floating-point form, you must force Maple to approximate it.

Right-click directly on the output expression just created. Choose the Approximate item and then 20 from the submenu. A call to evalf is automatically inserted directly below the symbolic output.

8. To find the value of the definite integral

$$\int_{\pi/4}^{\pi/3} x^2 \sin(x)\, dx = -x^2 \cos(x) + 2\cos(x) + 2x \sin(x) \Big|_{x=\pi/4}^{x=\pi/3},$$

you can evaluate answer at $x = \pi/3$ and $x = \pi/4$ and subtract the two results from one another. Place the cursor immediately to the left of the semicolon in the evaluation command. Then type in - eval( answer, x=Pi/4 ) and press enter to execute the command.

---

[2]For the remainder of this book, the term "right-click" will be used.

```
> eval( answer, x=Pi/3 ) - eval( answer, x=Pi/4 );
```

$$-\frac{1}{18}\pi^2 + 1 + \frac{1}{3}\pi\sqrt{3} + \frac{1}{32}\pi^2\sqrt{2} - \sqrt{2} - \frac{1}{4}\pi\sqrt{2}$$

The result of this computation overwrites the result that this region previously displayed, and Maple moves the cursor to the next input region.

9. Now you can introduce a symbolic parameter, $a$, in the integrand. Scroll up to the top of the worksheet, and click on and highlight the $x$ in the sin($x$) in the original input field. Notice that you are automatically toggled back to *standard math* mode and that the *edit field* is filled with x. Type - a and hit *enter* twice, once to alter the integrand and once to recompute the input.

   Recompute the integral by right-clicking on the new integrand and choosing Integrate and x. Notice that the automatically generated input and output fields are *appended* in front of the previously computed integral. To remove the outdated material, click on the bracket to its left and then hit *Ctrl-Delete*.

   Reassign the new result to answer by placing the cursor in the answer := % input and hitting *enter*. Now all the necessary values are updated.

10. To investigate the answer further, you must first insert a new prompt. Move the cursor either to the answer:= % input region or to the corresponding output region. Then click on the ▷ button. Again, toggle this new input field to command-line mode.

11. An important aspect of exploring a mathematical problem is to visualize it. The variable, answer, depends on two variables, $x$ and $a$, so you can plot it as a surface in three-dimensional space. Type in the following command at the new prompt and press enter. Your cursor might change to a clock for a couple of seconds while Maple generates the plot.

```
> plot3d(answer, x=-Pi..Pi, a=0..1);
```

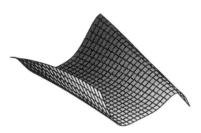

The displayed surface provides a concise representation of the effect on the value of the integral when you vary the parameter $a$.

There are several other ways to create plots in your worksheet, including using context menus and "drag and drop" operations.

12. You can modify many of the details of how Maple displays a plot. For example, you might want to add axes to the plot of the surface above. First, click on the plot so that the menu bar and context bar change to those specific to plotting. See figure 1.3. Then choose Boxed from the Axes menu. If your platform does not automatically redraw the plot with boxed axes, double-click on the plot to redisplay the surface.

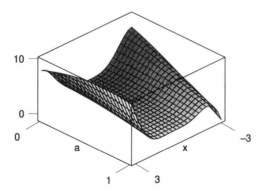

13. Maple can draw the surface in several styles. Right-click on the plot to bring up the context menu and choose Patch and contour from the Style menu. You might also want to change the viewing angle. The context bar displays the viewing angle as two parameters, $\theta$ and $\phi$. You can change the viewing angle either by editing the values for the two

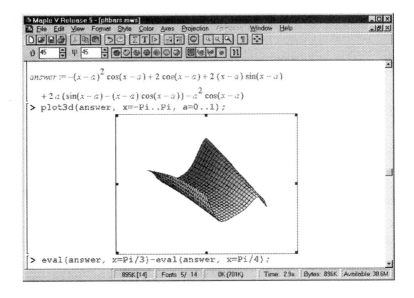

**FIGURE 1.3**    THE MENU AND CONTEXT BARS WHEN PLOTTING

parameters directly or by rotating the plot interactively: drag it around with the mouse. Try rotating the plot until $\theta = 120$ and $\phi = 45$.

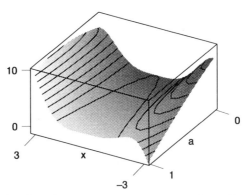

14. You can also display the answer as an animation. Use the Insert New Prompt button to insert a prompt below the graph. The animate command is part of the plots package, so you must load that package before you use animate.

First right-click on the input field and toggle off the Standard Math option. You can end commands with a colon instead of a semicolon when you don't wish to see the output.

```
> with(plots):
```

The cursor is now at the beginning of the next input region. To insert a new input region directly after the call to `with`, choose `Before Cursor` from the `Execution Group` item of the `Insert` menu. Then type in the following at the new prompt.

```
> animate(answer, x=-Pi..Pi, a=0..1);
```

To start the animation, click on the new plot so that the menu bar and context bar change to those specific for animations. See figure 1.4. Then click on the ▶ button.

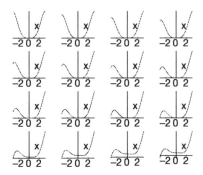

15. Maple can display animations in several ways. If you click on the ↻ button and restart the animation by clicking on the ▶ button, the animation keeps running until you click on the ■ button. You can control the speed with the ◀◀ and ▶▶ buttons.

16. The two evaluation commands in your worksheet below the plot are no longer relevant, so you might want to delete them. Do this for each input/output grouping by clicking on the bracket to its left and hitting Ctrl-Delete.

17. Save the current state of this worksheet by choosing `Save` from the `File` menu. Choose your own name for the file, but remember it for later; we will be reusing this worksheet in tutorials 3 and 4.

## 1.3 Tutorial 2: Managing Expressions through the Worksheet Interface

Besides using command-line syntax, there are several other ways of managing expressions that are accessed more directly through the worksheet

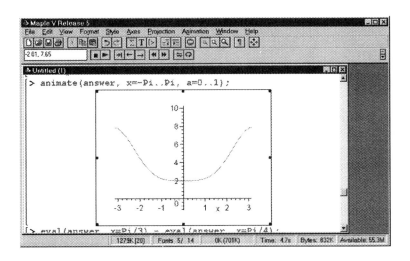

**FIGURE 1.4**   THE MENU AND CONTEXT BARS SPECIFIC TO ANIMATIONS

interface. Included among these are the use of spreadsheets, palettes, and drag and drop operations.

This tutorial illustrates the following key interface concepts:

- how to create and use spreadsheets
- how to use palettes
- how to take advantage of "smart plots"
- how to use context menus

## Spreadsheets

Spreadsheets are not available on all platforms. If your platform does not support them, please skip to the *Palettes* section.

First you need to restart Maple, creating a new session and an empty worksheet.

1. Create a new *spreadsheet* by selecting the Spreadsheet item from the Insert menu. An empty spreadsheet is created, with its upper leftmost cell (denoted *A1*) highlighted. See figure 1.5.

2. Type the character n. Notice that n appears in the edit field in the context bar. Also notice that the *A1* cell has a cross-hatched background, indicating that it is currently being defined.

    Press *enter* and cell *A1* is filled in.

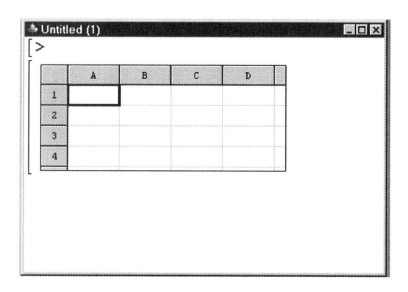

**FIGURE 1.5**   A MAPLE SPREADSHEET

3. You're going to want to see more rows and columns in your spreadsheet. Click just over the top of the spreadsheet so that you get a drag box outline around it. See figure 1.6. Click on the small black square at the bottom right and drag outwards until the columns read to *G* and the rows to *10*.

4. Now complete the first column. In cell *A2*, enter the value 1. Now select the cells *A2* through *A8* by first clicking on *A2* and then clicking on *A8* while holding down the *shift* key. See figure 1.7.

   Click on the ▤↓ button on the context bar. In the dialog box that appears, select *Down* and a *Step Size* of 1. Notice that cells *A3* through *A8* are filled in with the values 2 through 7, respectively.

5. You can also fill in cells with values that are tied to the values of other cells. Enter the formula `euler(~A1, x)` in cell *B1*. The ~A1 represents the *value* in cell *A1*. Notice that *B1* now contains `euler(n, x)`.

   Now select cells *B1* through *B8*. Select the `Fill` item from the `Spreadsheet` menu and then select the `Down` item from the submenu that appears. The seven other cells are automatically filled in with the first seven Euler polynomials. See figure 1.8.

   You may notice that, in filling in column *B*, some of the rows and columns previously visible have been "lost". Expand the worksheet window to fill the worksheet space by clicking on the expand button

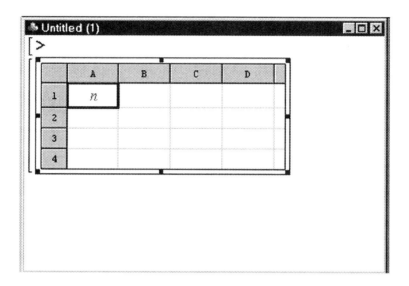

**FIGURE 1.6**    SPREADSHEET DRAG BOX

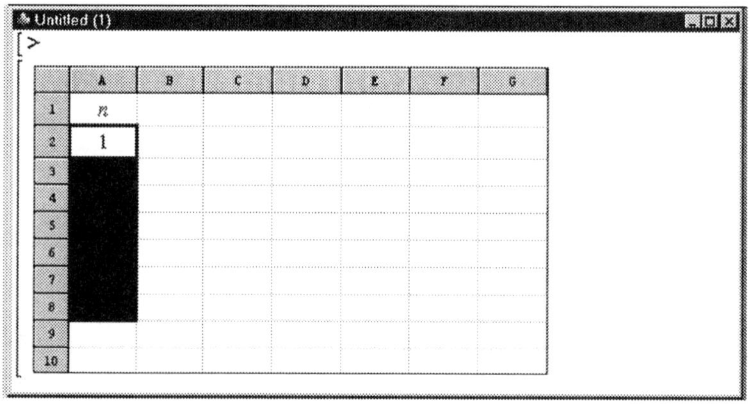

**FIGURE 1.7**    FILLING A COLUMN

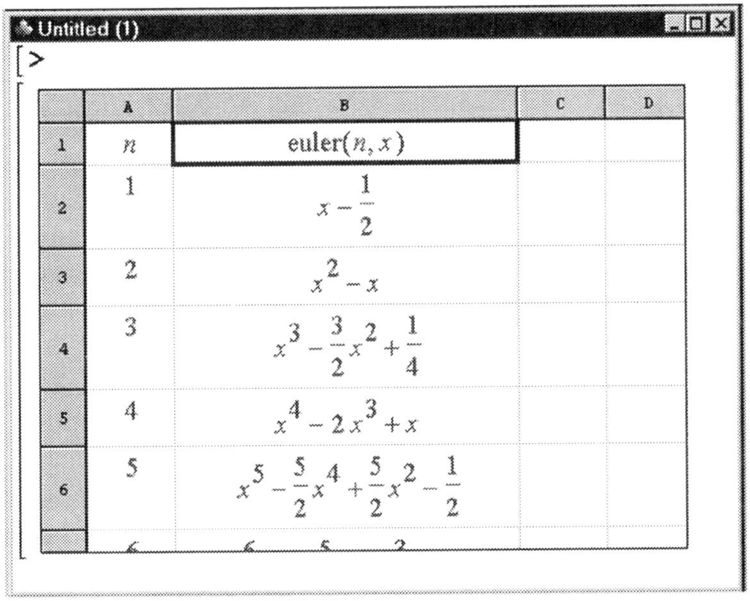

**FIGURE 1.8**    CELL DEPENDENCIES

in the upper-right corner of the window. Now expand the spreadsheet itself, as you did previously.

6. In a similar manner to column *B*, fill column *C* with the first derivatives of the Euler polynomials in column *B*. (Hint: define cell *C1* to be `diff(~B1, x)`).

   As columns can be filled in, so can rows.

7. Click on cell *A8* and enter a new value, 9, to replace 7. Notice that the two other cells in row *8* are automatically "crossed out". See figure 1.9. This is to indicate that they are no longer in agreement with *A8*, that they have gone "stale". Click on the ▤! button on the context bar to recompute all stale cells.

## Palettes

As you saw in tutorial 1, the worksheet interface includes *palettes*, blocks of predefined templates that help you create expressions, commands, special characters, and matrices. The following steps will show you how to use the expression palette as well as introduce you to some of the `Cut/Copy/Paste` and "drag and drop" options available in the worksheet interface.

1. Create a new input field beneath the spreadsheet by clicking on the ▷ button. If the new input field is in command-line mode, right-click on it

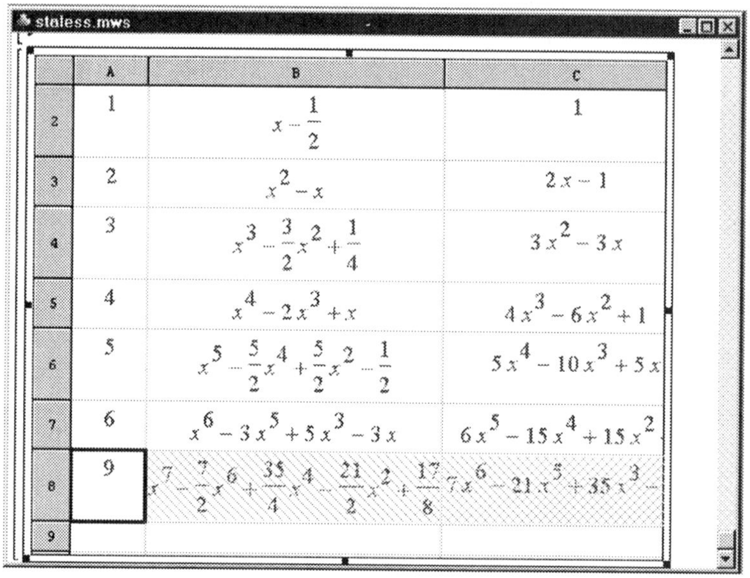

**FIGURE 1.9**   CELLS THAT NEED REFRESHING

and toggle it back to Standard Math mode. Go back to the spreadsheet and select cell *B7* (which should contain $x^6 - 3x^5 + 5x^3 - 3x$). Choose Copy from the Edit menu to place a copy of the expression on the clipboard.

2. Place the cursor back at the new input prompt and select Paste from the Edit menu.

3. Highlight this expression by dragging across it with the left mouse button so that it is entirely "blacked out". If the palettes are not already on your screen, open the expression palette by selecting Palettes from the View menu and choosing Expression Palette from the submenu that appears.

4. Create an indefinite integral from the highlighted expression by clicking on the ∫ₐ template of the expression palette. The input field that is generated has the highlighted expression filled in where the *a* was in the template and has a selected ? where the variable of integration is to go. Type x and press *enter* to fill in this element of the integration. Press *enter* again to send the input to Maple's computational engine.

5. Select the most recent output (that is, $1/7\,x^7 - 1/2\,x^6 + 5/4\,x^4 - 3/2\,x^2$) by dragging across it with the mouse. Click on the ∫ₐ template of the expression palette. Fill in each of the three ? fields with appropriate values (for example, 1, 6, and x) and press *enter*. Then press *enter* once

more to compute the result. Evaluate this expression numerically by right-clicking and choosing `Approximate` and 10.

### Smart Plots

Another interesting interface tool is *smart plots*, which can be created by right-clicking on existing expressions. They allow several manipulations not offered in standard plots created by command syntax.

   Smart plots are not available on all platforms. If your platform does not support them, please skip to tutorial 3.

1. Right-click on the output to the integration command in the worksheet (that is, $1/7\,x^7 - 1/2\,x^6 + 5/4\,x^4 - 3/2\,x^2$) Select `Plots` from the context menu and `2D-Plot` from the submenu. A `smartplot` command is automatically generated and a plot is created with an $x$-range of $-10..10$.

2. Decrease the $x$-range of the plot by right-clicking on it and choosing the `Ranges` option of the `Axes` item. Change the *Horizontal Axis* values to -4.0 and 4.0, leave the *Vertical Axis* as *Default*, and click on *OK*.

3. Scroll the worksheet until both the smart plot and the preceding integral are visible at the same time. Highlight the integrand in the input field (that is, $x^6 - 3x^5 + 5x^3 - 3x$) and *Ctrl*-drag that area (notice the small box that travels with the cursor) and drop it on top of the existing smart plot. The new function is added to the old function, and the vertical axis automatically changes to incorporate both functions.

4. Now alter the original function in the smart plot. Place the cursor somewhere on the original lineplot (below the $x$-axis), at a point that is completely separate from the second lineplot. Notice that it becomes "grayed out". Right-click and choose the `Style` option from the context menu that appears. Select `Line Style` and `Dot`. See figure 1.11.

   A function can also be removed from smart plots by clicking on its graphical representation and *Ctrl*-dragging it outside the plot boundaries. The corresponding expression is inserted in the worksheet.

5. Three-dimensional smart plots can also be created. Using the skills learned earlier, go to the bottom of the worksheet, create the expression $x^2 \sin(y)$ and create a smart plot from it. Change the axes ranges and some rendering options.

## 1.4   Tutorial 3: Documenting Your Work

Whereas the first two tutorials focused on manipulating Maple commands and its interface in order to solve problems, this tutorial focuses on how to

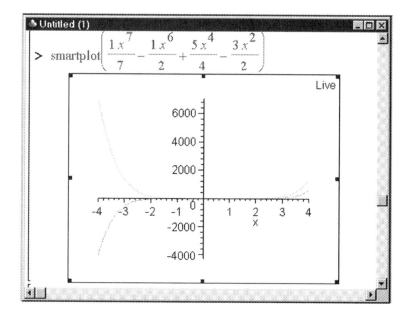

**FIGURE 1.11** SMART PLOTS

take the worksheet you created in the first tutorial and turn it into a small report. Open that worksheet by using the Open item from the File menu.

This tutorial illustrates the following key interface concepts.

- How to add text to your document.
- How to add formating using styles.
- How to place in-line mathematics in your text.

## Adding a Title

1. Start by giving your worksheet a title. The first step is to insert a new paragraph at the very top of your worksheet. Click on the bracket which contains the command defining the integrand $x^2 \sin(x)$.

2. Choose Execution Group from the Insert menu and Before Cursor from the ensuing submenu.

3. Since you wish to create a text region, not a command, click the **T** button in the tool bar.

4. You are now ready to type in the title. Type "An Indefinite Integral". The words will appear in the text region. At this stage they will look like normal text rather than a proper title.

5. To change how Maple typesets the title, you need to change its *style*. A style identifies a portion of text as being of a particular type. In addition,

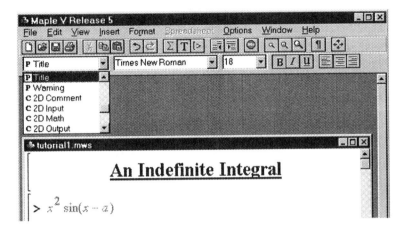

**FIGURE 1.12**  USING THE TITLE STYLE

it indicates how the text should appear in the document. Maple includes several predefined styles, including one for titles. Right now your text is typeset in the normal text style. The style droplist at the left-hand side of the context bar verifies this; it contains the word *Normal*. To change the style of your title, choose *Title* from the droplist. The words "An Indefinite Integral" will change in appearance. See figure 1.12.

6. Put your name as the author of this document. Place the cursor at the end of the title, then press enter. Type a name; for example, "by Jane Maplefan".

7. Notice that your name is in a different style than the title. Maple automatically formats the line directly after a *Title* field in *Author* style (as you can see from the current status of the style droplist).

## Adding Headings

The next step in constructing your document is to divide the commands in your worksheet into different sections.

1. The first Maple input regions should contain the commands that find the analytic answer: with the mouse, highlight the three input regions that define the integrand, compute the integral, and assign answer along with the corresponding output regions. Then choose Indent from the

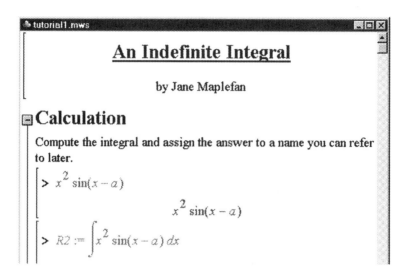

**FIGURE 1.13** MAKING A NEW SECTION

Format menu. A large bracket topped by a little square should appear to the left of the two commands that you highlighted.

2. Place the cursor in the text region immediately above the command that defines the integrand. Type in the heading, "Calculation". Then press enter. When you do so, the cursor moves to the next line. Since normal text usually follows section headings, Maple automatically assigns the *Normal* style to this new line of text.

3. Type the following brief introduction to the first command:

   Compute the integral and assign the answer to a name you can refer to later.

   See if your screen looks like figure 1.13.

4. Now you should put some text after the first command. Place the cursor in the output of the first command. Press the ▷ button to generate a new region and then the **T** button to turn the region into a text region.

5. Place the cursor in the new text region and type the following:

   The value of the integral is an anti-derivative of the integrand.

6. Write a conclusion to the first section. Place the cursor in the output region of the third (and last) command in the section, press the ▷ and **T** buttons as above, and type the following

As you can see, the answer depends on the parameter.

7. Create a section containing the calls to plot3d, with and animate. Use "Visualization" as the section heading and type the following as the text before the plot3d command.

> To see how the parameter affects the answer, you can graph it as a surface.

8. Type the following text after the plot of the surface:

> If the parameter represents time, you get the animation below.

### In-line Mathematics

Your worksheet needs a paragraph introducing the problem. For example, the following simple introduction

> Look at the integral $\int x^2 \sin(x-a)\,dx$. Notice that its integrand, $x^2 \sin(x-a)$, depends on the parameter $a$.

1. Place the cursor at the end of the word "Calculation." Press enter and type "Look at the integral ".

2. The next piece of your introduction is mathematical, so choose Maple Input from the Insert menu. Then type the Maple code corresponding to $\int x^2 \sin(x-a)\,dx$; that is, int(x^2 * sin(x-a), x).[3] Look at the *edit field* in the context bar. See figure 1.14. It shows the code you are typing. Press *enter*. The worksheet now displays the integral using standard mathematical notation.

3. Choose the Text Input item from the Insert menu to leave Maple Input and return to plain text mode. Type in the appropriate text following the integral. You should insert the $x^2 \sin(x - a)$ and the $a$ as mathematics in the same way you inserted the integral.

> Your document is complete. Save it and print it.

## 1.5  Tutorial 4: Multiple Worksheets

Maple allows you to work with several worksheets simultaneously. In this tutorial you will create a second worksheet, partially by dragging and dropping material from the worksheet you completed in the previous tutorial.

This tutorial illustrates the following key interface concepts.

---

[3] Alternatively, you can create this expression using the expression palette, as seen in tutorial 1.

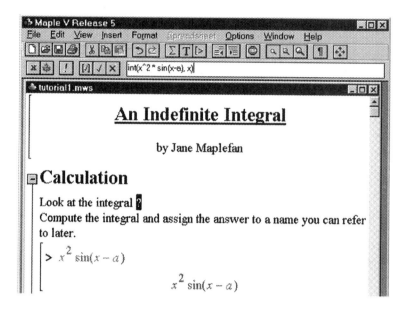

**FIGURE 1.14** THE EDIT FIELD

- How to perform drag and drop operations.
- How to insert hyperlinks.
- How to insert and use bookmarks.

## Drag and Drop

First you need to open the worksheet you made in tutorial 3 along with
a new blank worksheet. Choose `Vertical` from the `Window` menu so that
you can conveniently see both worksheets.

1. Place the cursor in the old worksheet in the command that performs
   the integration (that is, the one that starts with R2 :=). Press enter. Re-
   executing the command ensures that the Maple kernel knows the value
   of the integral.

2. The cursor is now at the beginning of the command that defines `answer`.
   Press enter again.

```
> answer := % ;
```

$$answer := -(x - a)^2 \cos(x - a) + 2\cos(x - a)$$
$$+ 2(x - a)\sin(x - a)$$
$$+ 2a(\sin(x - a) - (x - a)\cos(x - a)) - a^2\cos(x - a)$$

You do not have to re-execute the plot3d and animate commands.

3. Place the cursor in the input region of the new worksheet.

4. Click on the **T** button and change the style to *Title*. Type in the title "Anti-Derivatives" and press enter twice.

5. Type in the following:

In the worksheet, An Indefinite Integral, Maple determines that

6. Click on the **⊳** button to insert a new prompt. Type the following command at the new prompt and press enter twice.

```
> answer;
```

$$-(x-a)^2 \cos(x-a) + 2\cos(x-a) + 2(x-a)\sin(x-a)$$
$$+ 2a(\sin(x-a) - (x-a)\cos(x-a)) - a^2 \cos(x-a)$$

7. Click on the **T** button and continue the text as follows: "is an anti-derivative of ".

8. The input region of the first command in the worksheet from tutorial 2 contains the integrand, namely $x^2 \sin(x-a)$. With the mouse, highlight this integrand.

9. Drag a *copy* of the expression over to the other worksheet by hitting *Ctrl* while dragging it with the mouse. Drop it at the end of the line you just typed.

10. Choose Text Input from the Insert menu to toggle back to plain text mode and continue entering the text as follows:

If you add a constant to the anti-derivative above, you get another anti-derivative.

11. Click on the **⊳** button to insert a new prompt. Type the following command at the new prompt.

```
> answer + C;
```

$$-(x-a)^2 \cos(x-a) + 2\cos(x-a) + 2(x-a)\sin(x-a)$$
$$+ 2a(\sin(x-a) - (x-a)\cos(x-a)) - a^2 \cos(x-a)$$
$$+ C$$

12. Click on the **T** button and type in the following text:

If you differentiate the above expression and simplify the result, you obtain the original integrand.

13. Right-click on the previous output and choose Differentiate and x from the context menu. Simplify that expression by highlighting it, right-clicking, and choosing Simplify.

14. Click on the 🖫 button to save the new worksheet.

### Adding Hyperlinks

Now that you have two worksheets, you can add hyperlinks between them. A *hyperlink* is a piece of text that, when you click on it, takes you to another location in the same or another worksheet, or to a help page.

1. In the original worksheet, highlight the words "an anti-derivative". You will find them just below the command that defines the integrand.

2. Choose Convert to from the Format menu and Hyperlink from the ensuing submenu.

3. A dialog box appears. Maple has already filled in the Link Text field with the highlighted text. In the Worksheet field, type in the name of the file that contains the new worksheet.

4. Click on the *OK* button to complete the hyperlink.

### Bookmarks

You can insert *bookmarks* in your worksheets to help you find certain locations in your worksheet. If you click on a hyperlink to a bookmark, Maple takes you to the location of the bookmark rather than to the top of the worksheet.

1. Place the cursor in the command that defines the integrand in the worksheet entitled *An Indefinite Integral*.

2. Choose Bookmarks from the View menu and Edit Bookmark from the submenu that appears.

3. A dialog box appears. Type in "integrand command" as the bookmark text. Click on the *OK* button to insert the new bookmark.

4. Move the cursor to the end of the worksheet. Choose Bookmarks from the View menu and integrand command from the submenu that appears. The cursor is now back at the location of your bookmark.

5. Click on the 🖫 button to save your worksheet with the bookmark in it.

6. In the worksheet entitled *Anti-derivatives*, highlight the words "An Indefinite Integral". You will find them near the top of the worksheet.

7. Choose Convert to from the Format menu and Hyperlink from the ensuing submenu.

8. A dialog box appears. Maple has already filled in the Link Text field with the highlighted text. In the Worksheet field, type in the name of the file that contains the worksheet entitled *An Indefinite Integral.*

9. In the Book Mark field, type in "integrand command."

10.Click on the *OK* button to insert the hyperlink.

Try out the hyperlinks. Close one of the worksheets and bring it back by clicking on the hyperlink in the other worksheet.

## 1.6   Tutorial 5: Getting Help

The previous sections provide a taste of the problem-solving and document preparation abilities of Maple V. While they serve as an introduction to these facilities, a description of all the features cannot possibly fit in this book. Maple contains literally thousands of commands, from the most general to the esoteric. The following chapters teach you the most useful and effective methods for finding some of them, but like this chapter, they are only an introduction. Maple does, however, come with its own complete, interactive reference manual, the Maple on-line help system. The help system allows you to explore Maple's commands and features by name or by subject. In addition, it allows you to locate help pages which contain a word or phrase of your choice. Since hyperlinks connect the help pages together, you can easily refer to related pages.

This tutorial introduces you to Maple V's help system so that you can obtain information whenever you need it. In particular, it presents the following key ideas.

- How to start up the help system.
- How to locate a page by subject.
- How to use the help browser.
- How to use the index.
- How to use the full text search facility.

Maple gives you three methods for locating information in the help system: by content, by topic searching, or by full text searching.

### The Contents of the Help System

You can find information in the help system much the same way you would find information in a book—by looking at the table of contents, which is provided in a help browser format.

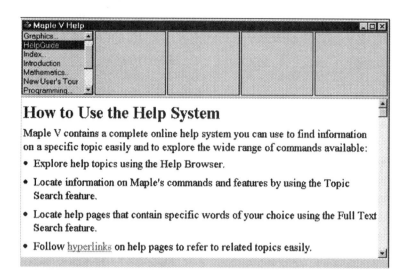

## How to Use the Help System

Maple V contains a complete online help system you can use to find information on a specific topic easily and to explore the wide range of commands available:

- Explore help topics using the Help Browser.

- Locate information on Maple's commands and features by using the Topic Search feature.

- Locate help pages that contain specific words of your choice using the Full Text Search feature.

- Follow hyperlinks on help pages to refer to related topics easily.

**FIGURE 1.15**    A HELP WINDOW

1. Choose `Using Help` from the `Help` menu. The help page that appears in the help window gives a brief introduction to the help system. The top section of the help window contains a hierarchical help browser for all of Maple's help pages. See figure 1.15.

2. Click on the words *Worksheet Interface* in the help browser (found by scrolling down in the leftmost column) and then `Introduction` in the subtopics that appear. A new help page appears in the window. Scroll down the help page until you find the title *Reference Material* and click on the + button to expand it. Click on the hyperlink called *help page menus*.

3. Open the *Help* item and read more about the items available in the `Help` menu.

### Searching by Topic

Sometimes you *almost* know the name of a command. For example, many programming languages use `sqr`, `sqrt`, or `root` to denote the square root function. You can search the help system by topic to find out which notation Maple uses.

1. Choose `Topic Search` from the `Help` menu.

2. The dialog box that appears allows you to type in the topic in which you are interested. If you have searched the help system by topic previously in the current session, the dialog box reflects the results of your last search.

3. Type *sq* into the Topic field. Maple displays all the help topics that begin with *sq*.

4. You can select any of the matching topics. Double-click on sqrt to see the help page on Maple's square root function.

Finding information in the help system by searching by topic is similar to finding information in this book by searching its index.

If you happen to already know the topic of the help page you want to read, you can conveniently access the help page directly from the worksheet: type *?topic* at the prompt and press enter to access the help page on *topic*. For example, the following command brings up the help page on the sin function (and the other trigonometric functions).

```
> ?sin
```

## Full Text Searching

When you search for information, you may know a few key words that should be on the help page you are looking for. For example, you can expect the help pages that explain how Maple helps you solve differential equations to contain the words *solve*, *differential*, and *equation*. You can search the whole help system for those words.

1. Choose Full Text Search from the Help menu.

2. The dialog box that appears allows you to type in the words for which you want to search. Type in *solve differential equation* in the Word(s) field. Then click on the *Search* button to tell Maple to start searching.

3. Maple lists all the matching topics along with numbers indicating how good the match is. You are likely to find the most interesting topics near the top. See figure 1.16.

4. You can select any of the matching topics. Double-click on dsolve to see the help page on Maple's primary command for solving ordinary differential equations.

As you work your way through the chapters in this book, feel free to access the wealth of information that is stored in the help system. With only one or two simple commands, you can locate the information you need on commands, their options and uses, and on the features of the interface.

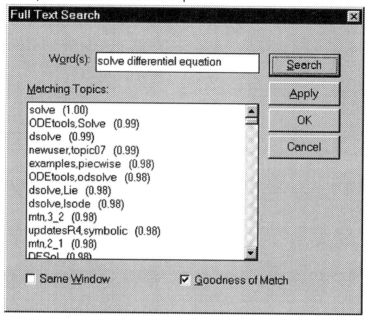

**FIGURE 1.16**    SEARCHING FOR SUBTOPICS

## 1.7   Conclusion

After working through the five tutorials in this chapter, you should feel comfortable working with Maple's advanced user interface. *Tutorial 1: Solving Problems* on page 5 was an introduction to solving mathematical problems. It demonstrates Maple's ease of use and its inherent ability to organize problem solving in a clear and efficient manner. In *Tutorial 2: Managing Expressions through the Worksheet Interface* on page 11, you learned to create and manipulate data and expressions through Maple's worksheet interface, through the use of spreadsheets, palettes, and other powerful tools. In *Tutorial 3: Documenting Your Work* on page 17, you learned to easily incorporate text and standard mathematical notation into the same worksheet. You can even insert executable commands, called in-line mathematics, in the middle of your sentence structures. In *Tutorial 4: Multiple Worksheets* on page 21, you learned how to cut and paste. You can paste text as well as Maple input and output. You also learned how to insert hyperlinks and bookmarks. *Tutorial 5: Getting Help* on page 25 reviewed the three methods for obtaining information through Maple's interactive help system: by content, by topic, or by full-text searching.

With these skills in hand, you can find many ways to manipulate and visualize complicated mathematical information and to create sophisticated technical documents.

# Mathematics with Maple: the Basics

This chapter begins with a discussion of numeric calculations in Maple, which differ slightly from most other mathematical applications. Basic symbolic computations and assignment statements follow. The final two sections teach the basic types of objects in Maple, and provide an introduction to the manipulation of objects and the commands most useful for this purpose.

You will get the most out of this book by using your computer to try out the examples as you read along. This chapter sketches out the Maple commands necessary to get you started. Subsequent chapters will give these and other commands a more in-depth treatment.

If you want to have a deeper understanding of what lies beneath the surface, use the on-line help facility. To obtain help about a command, simply type a question mark and then the name of the command, or in some cases just the topic you wish to learn more about, at the Maple prompt.

$$?command$$

## 2.1   Introduction

Throughout the remainder of this book, the command-line input format is used. To toggle all subsequent input fields to command-line mode, choose Maple Notation from the Input Display option of the Options menu. For information on how to use standard math (or two-dimensional) input mode, please refer to chapter 1.

The most basic computations in Maple are numerical. Maple can work like a conventional calculator with integers or floating-point numbers. Enter the expression using natural syntax. A semicolon signals the end of each calculation.

```
> 1 + 2;
```
$$3$$

```
> 1 + 3/2;
```
$$\frac{5}{2}$$

```
> 2*(3+1/3)/(5/3-4/5);
```
$$\frac{100}{13}$$

```
> 2.8754/2;
```
$$1.437700000$$

Of course, Maple can do much more, as you will see shortly.

For the moment, however, consider a simple example.

```
> 1 + 1/2;
```
$$\frac{3}{2}$$

Note that Maple performs exact calculations with rational numbers. The result of 1+1/2 is 3/2 not 1.5. To Maple, the rational number 3/2 and the floating-point approximation 1.5 are completely different objects. The ability to represent exact expressions allows Maple to preserve much more information about their origins and structure. The origin and structure of a number such as

$$.5235987758$$

are much less clear than for an exact quantity such as

$$\frac{1}{6}\pi$$

When you begin to deal with more complex expressions the advantage is greater still.

Maple can work not only with rational numbers, but also with arbitrary expressions. It can manipulate integers, floating-point numbers, variables, sets, lists, sequences, polynomials over a ring, and many more mathematical constructs. In addition, Maple is also a complete programming language which contains procedures, tables, and other programming constructs.

## 2.2 Numerical Computations

### Integer Computations

Integer calculations are straightforward. Remember to terminate each command with a semicolon.

```
> 1 + 2;
```

$$3$$

```
> 75 - 3;
```

$$72$$

```
> 5*3;
```

$$15$$

```
> 120/2;
```

$$60$$

Maple can also work with arbitrarily large integers. The practical limit on integers is approximately 500 000 digits, depending mainly on the speed and resources of your computer. Maple has no trouble calculating large integers, counting the number of digits in a number, or factoring integers. This final computation may take some time, as the integers are particularly large.

```
> 100!;
```

$$9332621544394415268169922388562667004907\backslash$$
$$15968264381621468592963895217599993229\backslash$$
$$9156089414639761565182862536979208 2722\backslash$$
$$37582511852109168640000000000000000000\backslash$$
$$00000$$

```
> length(%);
```

$$158$$

This answer indicates the number of digits in the last example. The ditto operator, (%), is simply a shorthand notation identifying the previous result. To recall the second or third most previous result, use %% and %%%, respectively.

**TABLE 2.1**   Commands for Working with Integers

| | |
|---|---|
| abs | absolute value of an expression |
| factorial | factorial of an integer |
| iquo | quotient of integer division |
| irem | remainder of integer division |
| iroot | roots of integers |
| isqrt | square root of integers |
| max, min | maximum and minimum of a set of inputs |
| mod | modulo arithmetic |
| surd | finding real roots |

```
> ifactor(60);
```

$$(2)^2 \ (3) \ (5)$$

In addition to `ifactor`, Maple has many commands for working with integers, some of which allow for calculations of a greatest common divisor (gcd) of two integers, integer quotients and remainders, and primality tests. See the examples below, as well as table 2.1.

```
> igcd(123, 45);
```

$$3$$

```
> iquo(25,3);
```

$$8$$

```
> isprime(18002676583);
```

*true*

## Exact Arithmetic—Rationals, Irrationals, and Constants

An important Maple property is the ability to perform exact rational arithmetic; that is, to work with rational numbers (fractions) without having to reduce them to floating-point approximations.

```
> 1/2 + 1/3;
```

$$\frac{5}{6}$$

Maple handles the rational numbers and produces an exact result. The distinction between *exact* and *approximate* results is an important one. The ability to perform exact computations with computers enables you to solve a whole new range of problems, and sets products like Maple apart from their purely numerical cousins.

Maple can produce floating-point estimates if required. In fact, Maple can work with floating-point numbers with many thousands of digits, so producing accurate estimates of exact expressions introduces no difficulty.

```
> Pi;
```

$$\pi$$

```
> evalf(Pi, 100);
```

$$3.141592653589793238462643383279502884 1 \backslash$$

$$971693993751058209749445923078164062 86 \backslash$$

$$20899862803482534211 7068$$

Learning how Maple distinguishes between exact and floating-point representations of values is important.

Here is an example of a rational (exact) number.

```
> 1/3;
```

$$\frac{1}{3}$$

The following is its floating-point approximation (shown to ten digits, by default).

```
> evalf(%);
```

$$.3333333333$$

These results are not the same mathematically, nor are they at all the same in Maple.

*Whenever you enter a number in decimal form, Maple treats it as a floating-point approximation.* In fact, the presence of a decimal number in an expression will cause Maple to produce an approximate floating-point result, since it cannot be sure of producing an exact solution from even partially approximate data.

```
> 3/2*5;
```

$$\frac{15}{2}$$

```
> 1.5*5;
```

$$7.5$$

Thus, you should *use floating-point numbers only when you wish to restrict Maple to working with non-exact expressions.*

Maple makes entering exact quantities easy by using symbolic representation, like $\pi$, in contrast to 3.14. Maple treats irrational numbers as exact quantities. Here is how you represent the square root of two in Maple.

```
> sqrt(2);
```

$$\sqrt{2}$$

Here is another square root example.

```
> sqrt(3)^2;
```

$$3$$

Maple knows the standard mathematical constants, such as $\pi$ and the base of the natural logarithms, e. It works with them as exact quantities.

```
> Pi;
```

$$\pi$$

```
> sin(Pi);
```

$$0$$

To write the exponential function, you must use the function exp.

```
> exp(1);
```

$$e$$

```
> ln(exp(5));
```

$$5$$

Actually, the second last example may look confusing. Remember that when Maple is producing "typeset" real-math notation, it attempts to represent mathematical expressions as you might write them yourself. Thus, you enter $\pi$ as Pi and Maple displays it as $\pi$. Maple is case sensitive, so be sure to use proper capitalization when stating these constants. The names Pi and pi are not equivalent! For more information on Maple constants, type ?constants at the prompt.

## Floating-Point Approximations

Although Maple prefers to work with exact values, it can return a floating-point approximation up to about 500 000 digits in length whenever you require it, depending upon your computer's resources.

Ten or twenty accurate digits in floating-point numbers may seem adequate for almost any purpose, but two problems, in particular, severely limit the usefulness of such a system.

First, when subtracting two floating-point numbers of almost equal magnitude the difference's relative error may be very large: if each of the two original numbers is identical in their first seventeen (out of twenty) digits, their difference is a three-digit number known to only three digits! In this case, you would need to use almost forty digits in order to have twenty accurate digits in the answer.

Second, a result's mathematical form is more concise, compact, and convenient than its numerical value. For instance, an exponential function provides more information about the nature of a phenomenon than even very large numbers. An exact analytical description can also determine the behavior of a function when extrapolating to regions for which no data exists.

The `evalf` command converts an exact numerical expression to a floating-point number.

```
> evalf(Pi);
```

$$3.141592654$$

By default, Maple calculates the result using ten digits of accuracy, but you may specify any number of digits. Simply indicate the number after the numerical expression, using the following notation.

```
> evalf(Pi, 200);
```

$$3.14159265358979323846264338327950288441\backslash$$
$$97169399375105820974944592307816406286\backslash$$
$$20899862803482534211706798214808651328\backslash$$
$$23066470938446095505822317253594081284\backslash$$
$$81117450284102701938521105559644622948\backslash$$
$$9549303820$$

You can also force Maple to do all its computations with floating-point approximations by including at least one floating-point number in each expression. Floats are "contagious": if an expression contains even one floating-point number, Maple evaluates the entire expression using floating-point arithmetic.

```
> 1/3 + 1/4 + 1/5.3;
```

$$.7720125786$$

```
> sin(0.2);
```

$$.1986693308$$

While the optional second argument to `evalf` controls the number of floating-point digits for that particular calculation, the special variable `Digits` sets the number of floating-point digits for all subsequent calculations.

```
> Digits := 20;
```

$$Digits := 20$$

```
> sin(0.2);
```

$$.19866933079506121546$$

`Digits` is now set to twenty, which Maple then uses at each step in a calculation. Maple works like a calculator or an ordinary computer application, in this respect. Remember that when you evaluate a complicated numerical expression, errors can build up which might cause the final result to be accurate to less than the full twenty digits. In general, setting `Digits` to produce a given accuracy is not easy, as the final result depends on your particular question. Using larger values, however, usually gives you some indication. Maple is very accommodating if extreme floating-point accuracy is important in your work.

## Arithmetic with Special Numbers

Maple V can work with complex numbers. $I$ is Maple's default symbol for the square root of minus one, that is, $I = \sqrt{-1}$.

```
> (2 + 5*I) + (1 - I);
```

$$3 + 4\,I$$

```
> (1 + I)/(3 - 2*I);
```

$$\frac{1}{13} + \frac{5}{13}\,I$$

You can also work with other bases and number systems.

```
> convert(247, binary);
```

$$11110111$$

```
> convert(1023, hex);
```

$$`3FF`$$

```
> convert(17, base, 3);
```

$$[2, 2, 1]$$

Maple returns an integer base conversion as a list of digits; otherwise, a line of numbers, like 221, may be ambiguous, especially when dealing with large bases. Note that Maple lists the digits in order from least significant to most significant.

Maple also supports modulo arithmetic.

```
> 27 mod 4;
```

$$3$$

Symmetric and positive representations are both available.

```
> mods(27,4);
```

$$-1$$

```
> modp(27,4);
```

$$3$$

The default for the mod command is positive representation, but you can change this option (see the help page ?mod for details).

Maple can also work with Gaussian Integers. The GaussInt package has about thirty commands for working with these special numbers. Type ?GaussInt for more information about these commands.

## Mathematical Functions

Maple V knows all the standard mathematical functions (see table 2.2 for a partial list).

```
> sin(Pi/4);
```

$$\frac{1}{2}\sqrt{2}$$

```
> ln(1);
```

$$0$$

When Maple cannot find a simpler form, it leaves the expression as it is rather than convert it to an inexact form.

```
> ln(Pi);
```

$$\ln(\pi)$$

**TABLE 2.2**  Mathematical Functions that Maple Knows

| | |
|---|---|
| sin, cos, tan, etc. | trigonometric functions |
| sinh, cosh, tanh, etc. | hyperbolic trigonometric functions |
| arcsin, arccos, arctan, etc. | inverse trigonometric functions |
| exp | exponential function |
| ln | natural logarithmic function |
| log[10] | logarithmic function base 10 |
| sqrt | algebraic square root function |
| round | round to the nearest integer |
| trunc | truncate to the integer part |
| frac | fractional part |
| BesselI, BesselJ, BesselK, BesselY | Bessel functions |
| binomial | binomial function |
| erf, erfc | error complementary error functions |
| Heaviside | Heaviside step function |
| Dirac | Dirac delta function |
| MeijerG | Meijer $G$ function |
| Zeta | Riemann Zeta function |
| LegendreKc, LegendreEc, LegendrePic | Legendre's elliptic integrals |
| LegendreKc1, LegendreEc1, LegendrePic1 | |
| hypergeom | hypergeometric function |

## 2.3  Basic Symbolic Computations

Maple V knows how to work with mathematical unknowns, and expressions which contain them.

```
> (1 + x)^2;
```

$$(1 + x)^2$$

```
> (1 + x) + (3 - 2*x);
```

$$4 - x$$

Note that Maple automatically simplifies the second expression.

Maple V has hundreds of commands for working with symbolic expressions.

```
> expand((1 + x)^2);
```

$$1 + 2x + x^2$$

```
> factor(%);
```

$$(1 + x)^2$$

As mentioned in *Numerical Computations* on page 31, the ditto operator, %, is a shorthand notation for the result of the previous command.

```
> Diff(sin(x), x);
```

$$\frac{\partial}{\partial x} \sin(x)$$

```
> value(%);
```

$$\cos(x)$$

```
> Sum(n^2, n);
```

$$\sum_{n} n^2$$

```
> value(%);
```

$$\frac{1}{3} n^3 - \frac{1}{2} n^2 + \frac{1}{6} n$$

Divide one polynomial in $x$ by another.

```
> rem(x^3+x+1, x^2+x+1, x);
```

$$2 + x$$

Create a series.

```
> series(sin(x), x=0, 10);
```

$$x - \frac{1}{6} x^3 + \frac{1}{120} x^5 - \frac{1}{5040} x^7 + \frac{1}{362880} x^9 + O(x^{10})$$

All of the mathematical functions mentioned in the previous section also accept unknowns as arguments.

## 2.4 Assigning Names to Expressions

Using the ditto operator, or retyping a Maple expression every time you want to use it, is not always convenient, so Maple allows you to name an object. Use the following syntax for naming.

$$\boxed{name := expression;}$$

You may assign a name to *any* Maple expression.

```
> var := x;
```

$$var := x$$

```
> term := x*y;
```

$$term := x\,y$$

You can even give names to equations.

```
> eqs := x = y + 2;
```

$$eqs := x = y + 2$$

Maple names can contain any alphanumeric characters and under-scores, but they *cannot start with a number*. Also, avoid starting names with an underscore as Maple saves these names for internal classification. Valid Maple names include: `polynomial`, `test_data`, `RoOt_1OcUs_pLoT`, and `value2`. Examples of *invalid* Maple names are 2ndphase (because it begins with a number), and x&y (because & is not an alphanumeric charac-ter).

You can define your own functions using Maple's *arrow notation* (->). This also lets Maple know how to evaluate the function when it appears in Maple expressions. At this point, you can do simple graphing of the function using the `plot` command.

```
> f := x -> 2*x^2 -3*x +4;
```

$$f := x \rightarrow 2x^2 - 3x + 4$$

```
> plot (f(x), x= -5...5);
```

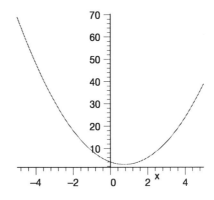

You can see more of the `plot` command in chapter 4.

The assignment (:=) operation can then associate a function name with the function definition. The name of the function is on the left-hand side of the :=. The function definition (using the arrow notation) is on the right-hand side. The following statement defines f as the "squaring function."

```
> f := x -> x^2;
```

$$f := x \rightarrow x^2$$

Then, evaluating f at an argument produces the square of f's argument.

```
> f(5);
```

$$25$$

```
> f(y+1);
```

$$(y + 1)^2$$

Not all names are available for variables. Maple has predefined and reserved a few. If you try to assign to a name which is predefined or reserved, Maple tells you that the name you have chosen is protected.

```
> Pi := 3.14;
```

```
Error, attempting to assign to 'Pi' which is protected
```

```
> set := {1, 2, 3};
```

```
Error,
attempting to assign to 'set' which is protected
```

## 2.5  More Basic Types of Maple Objects

Maple can be difficult to use without a brief introduction to other, more complex, types of objects it can represent. This section examines these basic types of Maple objects, including *expression sequences*, *lists*, *sets*, *arrays*, *tables*, and *strings*. These simple ideas are essential to the discussion in the rest of this book.

### Expression Sequences

The basic Maple data structure is the *expression sequence*. This is simply a group of Maple expressions separated by commas.

```
> 1, 2, 3, 4;
```

$$1, 2, 3, 4$$

```
> x, y, z, w;
```

$$x, y, z, w$$

Expression sequences are neither lists nor sets. They are a distinct data structure within Maple and have their own properties. For example, they preserve the order and repetition of their elements. Items will stay in the order you enter them and if you enter an element twice, both copies will remain. Other properties of sequences will become apparent as you progress through this manual. Sequences are often used to build more sophisticated objects through such operations as concatenation.

Sequences extend the capabilities of many basic Maple operations. For example, concatenation is a basic name-forming operation. The concatenation operator in Maple is " . " and you can use it in the following manner.

```
> a.b;
```

$$ab$$

When applying concatenation to a sequence, the operation affects each element. For example, if $S$ is a sequence then you can prepend the name a to each element in $S$ by concatenating a and $S$.

```
> S := 1, 2, 3, 4;
```

$$S := 1, 2, 3, 4$$

```
> a.S;
```

$$a1, a2, a3, a4$$

You can also perform multiple assignments using expression sequences. For example

```
> f,g,h := 3, 6, 1;
```

$$f, g, h := 3, 6, 1$$

```
> f;
```

$$3$$

```
> h;
```

$$1$$

## Lists

You create a *list* by enclosing any number of Maple objects (separated by commas) in square brackets.

```
> data_list := [1, 2, 3, 4, 5];
```

$$data\_list := [1, 2, 3, 4, 5]$$

```
> polynomials := [x^2+3, x^2+3*x-1, 2*x];
```

$$polynomials := [x^2 + 3, x^2 + 3x - 1, 2x]$$

```
> participants := [Kathy, Frank, Rene, Niklaus, Liz];
```

$$participants := [Kathy, Frank, Rene, Niklaus, Liz]$$

Thus, a list is an expression sequence enclosed in square brackets.

Maple preserves the order and repetition in a list. Thus, [a,b,c], [b,c,a], and [a,a,b,c,a] are all different.

> [a,b,c], [b,c,a], [a,a,b,c,a];

$$[a, b, c], [b, c, a], [a, a, b, c, a]$$

The fact that order is preserved allows you to extract a particular element of a list, without having to worry if it has been "moved".

> letters := [a,b,c];

$$letters := [a, b, c]$$

> letters[2];

$$b$$

Use the nops command to determine the total number of elements within a list.

> nops(letters);

$$3$$

You can also convert a list to an expression sequence using the op command.

> op(letters);

$$a, b, c$$

*The* nops *and* op *Commands* on page 57 discusses these two commands and their other uses in more detail.

## Sets

Maple supports *sets* in the mathematical sense. Commas separate the objects, as they do in a sequence or list, but the enclosing curly brackets identify the object as a set.

> data_set := {1, -1, 0, 10, 2};

$$data\_set := \{0, -1, 1, 2, 10\}$$

> unknowns := {x, y, z};

$$unknowns := \{x, z, y\}$$

Thus, a set is an expression sequence enclosed in curly brackets.

Maple does not preserve order or repetition in a set. That is, Maple sets have the same properties as sets do in mathematics. Thus, the following three sets are identical.

```
> {a,b,c}, {c,b,a}, {a,a,b,c,a};
```

$$\{b,\ c,\ a\},\ \{b,\ c,\ a\},\ \{b,\ c,\ a\}$$

Remember that to Maple the integer 2 is distinct from the floating-point approximation 2.0. Thus, the following set has three elements, not two.

```
> {1, 2, 2.0};
```

$$\{1,\ 2,\ 2.0\}$$

The properties of sets make them a particularly useful concept in Maple, just as they are in mathematics. Maple provides many operations on sets, including the basic operations of *intersection* and *union* using the notation intersect and union.

```
> {a,b,c} union {c,d,e};
```

$$\{e,\ b,\ c,\ a,\ d\}$$

```
> {1,2,3,a,b,c} intersect {0,1,y,a};
```

$$\{1,\ a\}$$

The nops command counts the number of elements of a set, as well as of a list.

```
> nops(%);
```

$$2$$

Another expression manipulator, op, mentioned in *Lists* on page 42, can also convert sets to expression sequences.

```
> op({1,2,3,a,b});
```

$$1,\ 2,\ 3,\ b,\ a$$

For more details, see *The* nops *and* op *Commands* on page 57.

A common and very useful command, often used on sets, is map. Mapping allows you to apply a function simultaneously to all the elements of any structure.

```
> numbers := {0, Pi/3, Pi/2, Pi};
```

$$numbers := \left\{\pi,\ 0,\ \frac{1}{3}\pi,\ \frac{1}{2}\pi\right\}$$

```
> map(g, numbers);
```

$$\left\{g(\pi),\ g(0),\ g\left(\frac{1}{3}\pi\right),\ g\left(\frac{1}{2}\pi\right)\right\}$$

```
> map(sin, numbers);
```

$$\left\{0,\ 1,\ \frac{1}{2}\sqrt{3}\right\}$$

Further examples demonstrating the use of map appear in *The* map *Command* on page 55 and *Mapping a Function onto a List or Set* on page 170.

## Operations on Sets and Lists

The member command validates membership in sets and lists.

```
> participants := [Kate, Tom, Steve];
```

$$participants := [Kate,\ Tom,\ Steve]$$

```
> member(Tom, participants);
```

*true*

```
> data_set := {5, 6, 3, 7};
```

$$data\_set := \{3,\ 5,\ 6,\ 7\}$$

```
> member(2, data_set);
```

*false*

To choose items from lists, use the subscript notation, [*n*], where *n* identifies the position of the desired element in the list.

```
> participants := [Kate, Tom, Steve];
```

$$participants := [Kate,\ Tom,\ Steve]$$

```
> participants[2];
```

*Tom*

Maple V understands *empty* sets and lists; that is, lists or sets which have no elements.

```
> empty_set := {};
```

$$empty\_set := \{\}$$

```
> empty_list := [];
```

$$empty\_list := []$$

You can create a new set from other sets by using, for example, the union command. Delete items from sets with minus.

```
> old_set := {2, 3, 4} union {};
```

$$old\_set := \{2,\ 3,\ 4\}$$

```
> new_set := old_set union {2, 5};
```

$$new\_set := \{2,\ 3,\ 4,\ 5\}$$

```
> third_set := old_set minus {2, 5};
```

$$third\_set := \{3,\ 4\}$$

## Arrays

*Arrays* are an extension of the concept of the list data structure. Think of a list as a group of items where you associate each item with a positive integer, its index, which represents its position in the list. The Maple array data structure is a generalization of this idea. Each element is still associated with an index, but is no longer restricted to one dimension. In addition, indices may also be zero or negative. Furthermore, you may define or change the array's individual elements without redefining it entirely.

Declare the array so Maple knows the dimensions you wish to use.

```
> squares := array(1..3);
```

$$squares := \text{array}(1..3,\ [])$$

Assign the array elements.

```
> squares[1] := 1;   squares[2] := 2^2;   squares[3] := 3^2;
```

$$squares_1 := 1$$
$$squares_2 := 4$$
$$squares_3 := 9$$

Or, if you prefer, do it all at once.

```
> cubes := array( 1..3, [1,8,27] );
```

$$cubes := [1,\ 8,\ 27]$$

You may select a single element using the same notation applied to lists.

```
> squares[2];
```

$$4$$

You must declare arrays in advance. To see the array's contents, you must use a command such as print.

```
> squares;
```

$$squares$$

```
> print(squares);
```

$$[1, 4, 9]$$

Having to say print explicitly may seem awkward at first. However, not only does Maple work more efficiently this way, but you can appreciate this feature if you work with large arrays.

The above array has only one dimension, but arrays in general can have more than one dimension. Define a 3 × 3 array.

```
> pwrs := array(1..3,1..3);
```

$$pwrs := \text{array}(1..3, 1..3, [])$$

This array has dimension two (two sets of indices). To begin, assign the array elements of the first row.

```
> pwrs[1,1] := 1;  pwrs[1,2] := 1;  pwrs[1,3] := 1;
```

$$pwrs_{1,1} := 1$$
$$pwrs_{1,2} := 1$$
$$pwrs_{1,3} := 1$$

Now continue for the rest of the array. If you prefer, you may end each command with a colon ( : ), instead of the usual semicolon ( ; ) to suppress the output.

```
> pwrs[2,1] := 2:  pwrs[2,2] := 4:  pwrs[2,3] := 8:
> pwrs[3,1] := 3:  pwrs[3,2] := 9:  pwrs[3,3] := 27:
> print(pwrs);
```

$$\begin{bmatrix} 1 & 1 & 1 \\ 2 & 4 & 8 \\ 3 & 9 & 27 \end{bmatrix}$$

You may select an element by specifying both the row and column.

```
> pwrs[2,3];
```

$$8$$

You can define a two-dimensional array and its elements all at once using a similar method employed for the one-dimensional example shown earlier. To do so, use lists within lists. That is, make a list where each element is a list that contains the elements of one row of the array. Thus, you could define the pwrs array as follows.

```
> pwrs2 := array( 1..3, 1..3, [[1,1,1], [2,4,8], [3,9,27]] );
```

$$pwrs2 := \begin{bmatrix} 1 & 1 & 1 \\ 2 & 4 & 8 \\ 3 & 9 & 27 \end{bmatrix}$$

Arrays are by no means limited to two dimensions, but those of higher order are more difficult to display. If you wish, you may also declare all the elements of the array at the same time as you make the definition.

```
> array3 := array( 1..2, 1..2, 1..2,
> [[[1,2],[3,4]], [[5,6],[7,8]]] );
```

$$array3 := \text{array}(1..2,\ 1..2,\ 1..2,\ [$$

$$(1,\ 1,\ 1) = 1$$

$$(1,\ 1,\ 2) = 2$$

$$(1,\ 2,\ 1) = 3$$

$$(1,\ 2,\ 2) = 4$$

$$(2,\ 1,\ 1) = 5$$

$$(2,\ 1,\ 2) = 6$$

$$(2,\ 2,\ 1) = 7$$

$$(2,\ 2,\ 2) = 8$$

$$])$$

Standard matrix operations and calculations are supported. Maple does not automatically expand the name of an array to the representation of all the elements. Thus, in some commands, you must specify explicitly that you wish to perform an operation on the elements.

Suppose that you wish to replace each occurrence of the number 2 in pwrs with the number 9. To do substitutions such as this, you can use the subs command. The basic syntax is

```
subs( x=expr1, y=expr2, ... , main_expr )
```

For example, suppose you wish to substitute $x + y$ for $z$ in an equation.

```
> expr := z^2 + 3;
```

$$expr := z^2 + 3$$

```
> subs( {z=x+y}, expr);
```

$$(x + y)^2 + 3$$

You might, however, be disappointed when the following call to subs does not work.

```
> subs( {2=9}, pwrs );
```

$$pwrs$$

You must instead force Maple to fully evaluate the name of the array to the component level and not just to its name, using the *evaluate matrix* command, evalm.

```
> subs( {2=9}, evalm(pwrs) );
```

$$\begin{bmatrix} 1 & 1 & 1 \\ 9 & 4 & 8 \\ 3 & 9 & 27 \end{bmatrix}$$

Not only does this cause the substitution to occur in the components as expected, but full evaluation also displays the array's elements, just as when you use the print command.

```
> evalm(pwrs);
```

$$\begin{bmatrix} 1 & 1 & 1 \\ 2 & 4 & 8 \\ 3 & 9 & 27 \end{bmatrix}$$

## Tables

A *table* is an extension of the concept of the array data structure. The difference between an array and a table is that a table can have *anything* for indices, not just integers.

```
> translate := table([one=un,two=deux,three=trois]);
```

$$translate := \text{table}([$$

$$two = deux$$

$$one = un$$

$$three = trois$$

$$])$$

```
> translate[two];
```

$$deux$$

Although at first they may seem to have little advantage over arrays, table structures are very powerful. Tables allow you to work with natural notation for data structures. For example, you can display the physical properties of materials using a Maple table.

```
> earth_data := table( [ mass=[5.976*10^24,kg],
>                              radius=[6.378164*10^6,m],
>                              circumference=[4.00752*10^7,m] ] );
```

$$earth\_data := \text{table}([$$

$$circumference = [.4007520000\,10^8,\, m]$$

$$mass = [.5976000000\,10^{25},\, kg]$$

$$radius = [.6378164000\,10^7,\, m]$$

$$])$$

```
> earth_data[mass];
```

$$[.5976000000\,10^{25},\, kg]$$

In this example, each index is a name and each entry is a list. In fact, this is a rather simple case. Often, much more general indices are useful. For example, you could construct a table which has algebraic formulæ for indices and the derivatives of these formulæ as values.

## Strings

A *string* is also an object in Maple, and is created by enclosing any number of characters in double quotes.

```
> "This is a string."
```

They are nearly indivisible units that stand only for themselves; they cannot be assigned a value.

```
> "my age" := 32;
```

```
on line 405, syntax error, unexpected string:
"my age" := 32;
              ^
```

Like elements of lists or arrays, the individual characters of a string can be indexed with square bracket notation.

```
> mystr := "I ate the whole thing.";
```

$$mystr := \text{“I ate the whole thing.”}$$

```
> mystr[11..-2];
```

$$\text{“whole thing”}$$

The concatenation operator, " . ", or the `cat` command is used to join two strings together, and the `length` command is used to determine the number of characters in a string.

```
> newstr := cat("I can't believe ", mystr);
```

$$newstr := \text{"I can't believe I ate the whole  thing."}$$

```
> length(newstr);
```

$$38$$

There are other commands that operate on strings and many more that takes strings as input.

## 2.6  Expression Manipulation

Many of Maple's commands concentrate on manipulating expressions. This includes manipulating results of Maple commands into a familiar form, or a form with which you want to work. This can also involve manipulating your own expressions into a form with which Maple can work. In this section we introduce the most commonly used commands in this area.

### The `simplify` Command

You can use this command to apply simplification rules to an expression. Maple has simplification rules for various types of expressions and forms, including trigonometric functions, radicals, logarithmic functions, exponential functions, powers, and various special functions.

```
> expr := cos(x)^5 + sin(x)^4 + 2*cos(x)^2
> - 2*sin(x)^2 - cos(2*x);
```

$$expr :=$$
$$\cos(x)^5 + \sin(x)^4 + 2\cos(x)^2 - 2\sin(x)^2 - \cos(2x)$$

```
> simplify(expr);
```

$$\cos(x)^5 + \cos(x)^4$$

To perform only a certain type of simplification, specify the type you desire.

```
> simplify(sin(x)^2 + ln(2*y) + cos(x)^2);
```

$$1 + \ln(2) + \ln(y)$$

```
> simplify(sin(x)^2 + ln(2*y) + cos(x)^2, 'trig');
```

$$1 + \ln(2\,y)$$

```
> simplify(sin(x)^2 + ln(2*y) + cos(x)^2, 'ln');
```

$$\sin(x)^2 + \ln(2) + \ln(y) + \cos(x)^2$$

With the *side relations* feature, you can also apply your own simplification rules. Indeed, you can program your own simplification rules by programming your own procedures, but that is beyond the scope of this book.

```
> siderel := {sin(x)^2 + cos(x)^2 = 1};
```

$$siderel := \{\sin(x)^2 + \cos(x)^2 = 1\}$$

```
> trig_expr := sin(x)^3 - sin(x)*cos(x)^2 + 3*cos(x)^3;
```

$$trig\_expr := \sin(x)^3 - \sin(x)\cos(x)^2 + 3\cos(x)^3$$

```
> simplify(trig_expr, siderel);
```

$$2\sin(x)^3 - 3\cos(x)\sin(x)^2 + 3\cos(x) - \sin(x)$$

## The factor Command

This command factors polynomial expressions.

```
> big_poly := x^5 - x^4 - 7*x^3 + x^2 + 6*x;
```

$$big\_poly := x^5 - x^4 - 7\,x^3 + x^2 + 6\,x$$

```
> factor(big_poly);
```

$$x\,(x-1)\,(x-3)\,(x+2)\,(x+1)$$

```
> rat_expr := (x^3 - y^3)/(x^4 - y^4);
```

$$rat\_expr := \frac{x^3 - y^3}{x^4 - y^4}$$

Both the numerator and denominator contain the common factor $x - y$; thus, factoring cancels these terms.

```
> factor(rat_expr);
```

$$\frac{x^2 + x\,y + y^2}{(y+x)\,(x^2+y^2)}$$

Maple can factor both univariate and multivariate polynomials over the domain the coefficients specify. You can also factor polynomials over algebraic extensions. See ?Factor for details.

## The expand Command

expand is essentially the reverse of factor. It causes the expansion of multiplied terms as well as a number of other expansions. This is among the most useful of the manipulation commands. Although you might imagine that with a name like expand the result would be larger and more complex than the original expression; this is not usually the case. In fact, expanding many expressions results in substantial simplification.

```
> expand((x+1)*(x+2));
```

$$x^2 + 3x + 2$$

```
> expand(sin(x+y));
```

$$\sin(y)\cos(x) + \cos(y)\sin(x)$$

```
> expand(exp(a+ln(b)));
```

$$e^a b$$

If you invoke expand with two (or more) arguments, it expands the first argument without expanding the given subexpressions.

```
> expand((x+1)*(y+z), x+1);
```

$$(x + 1)y + (x + 1)z$$

expand is quite flexible; not only can you specify that certain subexpressions be left alone during the expansion, but you can also program custom expansion rules.

simplify may seem like the most useful command when using Maple for the first time, but this is misleading. Unfortunately, the word *simplify* is rather vague. When you request to simplify an expression, Maple examines your expression, tests out many techniques, and then tries applying the appropriate simplification rules. However, this might take a little time. As well, Maple may not be able to guess what you want to accomplish since universal mathematical rules do not define what is simpler.

When you do know which manipulations will make your expression simpler for you, specify them directly. In particular, the expand command is among the most useful. It frequently results in substantial simplification, and also leaves expressions in a convenient form for many other commands.

## The convert Command

This command converts expressions between different forms.

**TABLE 2.3**   Common Conversions

| | |
|---|---|
| polynom | converting series to polynomials |
| exp, expln, expsincos | converting trigonometric expressions to exponential form |
| parfrac | converting rational expressions to partial fraction form |
| rational | converting floating-point numbers to rational form |
| radians, degrees | converting between degrees and radians |
| set, list, listlist | converting between data structures |

```
> convert(cos(x),exp);
```

$$\frac{1}{2} e^{(Ix)} + \frac{1}{2} \frac{1}{e^{(Ix)}}$$

```
> convert(1/2*exp(x) + 1/2*exp(-x),trig);
```

$$\cosh(x)$$

```
> A := array(1..2,1..2, [[a,b],[c,d]]);
```

$$A := \left[ \begin{array}{cc} a & b \\ c & d \end{array} \right]$$

```
> convert(A, 'listlist');
```

$$[[a, b], [c, d]]$$

```
> convert(A, 'set');
```

$$\{a, d, b, c\}$$

```
> convert(%, 'list');
```

$$[a, d, b, c]$$

## The normal Command

This command transforms rational expressions into *factored normal form*,

$$\frac{numerator}{denominator},$$

where the *numerator* and *denominator* are relatively prime polynomials with integer coefficients. That is, the normal command puts fractions on a common denominator.

```
> rat_expr_2 := (x^2 - y^2)/(x - y)^3 ;
```

$$rat\_expr\_2 := \frac{x^2 - y^2}{(-y + x)^3}$$

```
> normal(rat_expr_2);
```

$$\frac{y+x}{(-y+x)^2}$$

```
> normal(rat_expr_2, 'expanded');
```

$$\frac{y+x}{y^2-2\,x\,y+x^2}$$

The expanded option forces Maple to expand the polynomials in the numerator and the denominator.

## The combine Command

This command combines terms in sums, products, and powers into a single term. These transformations are, in some cases, the reverse of the transformations that expand applies.

```
> combine(exp(x)^2*exp(y),exp);
```

$$e^{(2x+y)}$$

```
> combine((x^a)^2, power);
```

$$x^{(2a)}$$

## The map Command

This command is most useful when working with lists, sets, and arrays. It provides an especially convenient means for working with multiple solutions or for applying an operation to each element of an array.

map applies a command to each element of a data structure or expression. While it is possible to write program structures such as loops to accomplish these tasks, you should not underestimate the convenience and power of the map command. map is one of the most useful commands in Maple. Take an extra minute to make sure you understand how to use this command.

```
> map( f, [a,b,c] );
```

$$[f(a),\ f(b),\ f(c)]$$

```
> data_list := [0, Pi/2, 3*Pi/2, 2*Pi];
```

$$data\_list := \left[0,\ \frac{1}{2}\,\pi,\ \frac{3}{2}\,\pi,\ 2\,\pi\right]$$

```
> map(sin, data_list);
```

$$[0,\ 1,\ -1,\ 0]$$

If you give the `map` command more than two arguments, Maple passes the last arguments to the initial command.

```
> map( f, [a,b,c], x, y );
```

$$[f(a,\ x,\ y),\ f(b,\ x,\ y),\ f(c,\ x,\ y)]$$

For example, to differentiate each item in a list with respect to $x$, you might use the following commands.

```
> fcn_list := [sin(x),ln(x),x^2];
```

$$fcn\_list := [\sin(x),\ \ln(x),\ x^2]$$

```
> map(Diff, fcn_list, x);
```

$$\left[ \frac{\partial}{\partial x}\sin(x),\ \frac{\partial}{\partial x}\ln(x),\ \frac{\partial}{\partial x}x^2 \right]$$

```
> map(value, %);
```

$$\left[ \cos(x),\ \frac{1}{x},\ 2x \right]$$

Not only can the procedure be an existing command, but you can also create an operation to map onto a list. For example, suppose that you wish to square each element of a list. Ask Maple to replace each element (represented by $x$) with its square $(x^2)$.

```
> map(x->x^2, [-1,0,1,2,3]);
```

$$[1,\ 0,\ 1,\ 4,\ 9]$$

## The `lhs` and `rhs` Commands

These two commands allow you to take the left-hand side and right-hand side of an expression, respectively.

```
> eqn1 := x+y=z+3;
```

$$eqn1 := y + x = z + 3$$

```
> lhs(eqn1);
```

$$y + x$$

```
> rhs(eqn1);
```

$$z + 3$$

## The `numer` and `denom` Commands

These two commands allow you to take the numerator and denominator of a rational expression, respectively.

```
> numer(3/4);
```

$$3$$

```
> denom(1/(1 + x));
```

$$x + 1$$

## The `nops` and `op` Commands

These two commands are useful for breaking expressions down into parts, and extracting subexpressions.

nops tells you how many parts an expression has. You may recall a discussion on these commands in *More Basic Types of Maple Objects* on page 41.

```
> nops(x^2);
```

$$2$$

```
> nops(x + y);
```

$$2$$

The op command allows you to access the parts of an expression. It returns the parts as a sequence.

```
> op(x^2);
```

$$x, 2$$

You can also ask for specific items by number or range.

```
> op(1, x^2);
```

$$x$$

```
> op(2, x^2);
```

$$2$$

```
> op(1..2, x+y+z+w);
```

$$x, y$$

## Common Questions about Expression Manipulation

### How do I Substitute for a Product of Two Unknowns?
Use side relations to specify an "identity." Substituting directly does not usually work.

```
> expr := a^3*b^2;
```

$$expr := a^3\, b^2$$

```
> subs(a*b=5, expr);
```

$$a^3\, b^2$$

Here the subs command was unsuccessful in its attempt to substitute. Try the simplify command this time to get the correct answer.

```
> simplify(expr, {a*b=5});
```

$$25\, a$$

**Why is the Result of** simplify **Not the Simplest Form?**   For example:

```
> expr2 := cos(x)*(sec(x)-cos(x));
```

$$expr2 := \cos(x)\,(\sec(x) - \cos(x))$$

```
> simplify(expr2);
```

$$1 - \cos(x)^2$$

The expected form was $\sin(x)^2$.

Again, use side relations to specify the identity.

```
> simplify(%, {1-cos(x)^2=sin(x)^2});
```

$$\sin(x)^2$$

The issue of simplification is a complicated one because of the difficulty of defining a "simple" form of an expression. One user's idea of a simple form may be vastly different from another user's; indeed, the idea of the simplest form can vary from situation to situation.

**How do I Factor the Constant out of** $2x + 2y$**?**   Currently, this operation is not possible in Maple because its simplifier automatically distributes the number over the product, believing that a sum is simpler than a product. In most cases, this is true.

```
> x^19 - x;
```

$$x^{19} - x$$

```
> factor(x^19 - x);
```

$$x\,(x - 1)\,(x^2 + x + 1)\,(x^6 + x^3 + 1)\,(x + 1)\,(1 - x + x^2)$$
$$(1 - x^3 + x^6)$$

If you enter the expression

```
> 2*(x + y);
```

$$2 x + 2 y$$

you see that Maple automatically multiplies the constant into the expression.

How can you then deal with such expressions, when you need to factor out constants, or negative signs? Should you need to factor such expressions, try this "clever" substitution.

```
> expr3 := 2*(x + y);
```

$$expr3 := 2 x + 2 y$$

```
> subs( 2=two, expr3 );
```

$$x \, two + y \, two$$

```
> factor(%);
```

$$two \, (x + y)$$

## 2.7 Conclusion

In this chapter you have seen many of the types of objects which Maple is capable of manipulating, including sequences, sets, and lists. You have seen a number of commands, including expand, factor, and simplify, which are most useful for manipulating and simplifying algebraic expressions. Others, such as map, are useful for sets, lists, and arrays. Meanwhile, subs is useful almost any time.

In the next chapter, you will learn to apply these concepts in order to tackle one of the most fundamental tasks in mathematics, the problem of solving systems of equations. As you learn about new commands, observe how the concepts of this chapter are used in setting up problems and manipulating solutions.

# Finding Solutions

This chapter introduces the key concepts needed for quick and concise problem solving in Maple. By learning how to use such tools as solve, map, subs, and unapply, you can save yourself a substantial amount of work. In addition, this chapter examines how all these commands interoperate.

## 3.1  Simple solve

Maple's solve command is a general-purpose equation solver. It takes a set of one or more equations and attempts to solve it exactly for the specified set of unknowns. (Recall from *Sets* on page 43 that you use braces to denote a set.) In the following examples you are solving one equation for one unknown, so each set contains only one element.

```
> solve({x^2=4}, {x});
```

$$\{x = 2\}, \{x = -2\}$$

```
> solve({a*x^2+b*x+c=0}, {x});
```

$$\left\{ x = \frac{1}{2}\frac{-b+\sqrt{b^2-4\,a\,c}}{a} \right\}, \left\{ x = \frac{1}{2}\frac{-b-\sqrt{b^2-4\,a\,c}}{a} \right\}$$

Maple returns each possible solution as a set. Since both of these equations have two solutions, Maple returns a sequence of solution sets. If you do not specify any unknowns in the equation, Maple solves for all of them.

```
> solve({x+y=0});
```

$$\{x = -y, \ y = y\}$$

Here you get only one solution set containing two equations. This result means that $y$ can take any value, while $x$ is the negative of $y$. This solution is parameterized with respect to $y$.

If you give an expression rather than an equation, Maple automatically assumes that the expression is equal to zero.

```
> solve({x^3-13*x+12}, {x});
```

$$\{x = 1\}, \ \{x = 3\}, \ \{x = -4\}$$

The `solve` command can also handle systems of equations.

```
> solve({x+2*y=3, y+1/x=1}, {x,y});
```

$$\{x = -1, \ y = 2\}, \ \left\{x = 2, \ y = \frac{1}{2}\right\}$$

Although you do not always need the braces (denoting a set) around either the equation or variable, using them forces Maple to return the solution as a set, which is usually the most convenient form. For example, the first thing you usually do with a solution is check it by substituting it back into your original equations. The following example demonstrates this procedure.

As a set of equations, the solution is in an ideal form for the `subs` command. You might first give the set of equations a name, like eqns, for instance.

```
> eqns := {x+2*y=3, y+1/x=1};
```

$$eqns := \left\{x + 2\,y = 3, \ y + \frac{1}{x} = 1\right\}$$

Then solve.

```
> soln := solve( eqns, {x,y} );
```

$$soln := \{x = -1, \ y = 2\}, \ \left\{x = 2, \ y = \frac{1}{2}\right\}$$

This produces two solutions: first,

```
> soln[1];
```

$$\{x = -1, \ y = 2\}$$

and second,

```
> soln[2];
```

$$\left\{ x = 2, \ y = \frac{1}{2} \right\}$$

## Verifying Solutions

To check the solutions, substitute them back into the original set of equations using the two-parameter eval command.

```
> eval( eqns, soln[1] );
```

$$\{1 = 1, \ 3 = 3\}$$

```
> eval( eqns, soln[2] );
```

$$\{1 = 1, \ 3 = 3\}$$

For verifying solutions, you will find that this method is generally the most convenient.

Observe that this application of the eval command has other uses. Suppose you wish to extract the value of $x$ from the first solution. Again, the best tool is the eval command.

```
> x1 := eval( x, soln[1] );
```

$$x1 := -1$$

Alternatively, you could extract the first solution for $y$.

```
> y1 := eval(y, soln[1]);
```

$$y1 := 2$$

You can even use this evaluation trick to convert solutions sets to other forms. For example, you could construct a list from the first solution where $x$ is the first element and $y$ is the second. First construct a list with the *variables* in the same order as you want the corresponding *solutions*.

```
> [x,y];
```

$$[x, \ y]$$

Then simply evaluate this list at the first solution.

```
> eval([x,y], soln[1]);
```

$$[-1, \ 2]$$

The first solution is now a list.

Instead, if you prefer that the solution for $y$ comes first, evaluate [y,x] at the solution.

```
> eval([y,x], soln[1]);
```

$$[2, -1]$$

Since Maple typically returns solutions in the form of sets (in which the order of objects is uncertain) remembering this method of extracting and working with solutions is useful.

map is another useful command which allows you to apply one operation to all solutions. For example, try substituting *both* solutions.

map applies the operation specified as its first argument to its second argument.

```
> map(f, [a,b,c], y, z);
```

$$[f(a, y, z), f(b, y, z), f(c, y, z)]$$

Due to the syntactical design of map, it cannot perform multiple function applications to sequences. Consider the previous solution sequence, for example.

```
> soln;
```

$$\{x = -1, y = 2\}, \left\{x = 2, y = \frac{1}{2}\right\}$$

Enclose soln in square brackets to convert it to a `list`.

```
> [soln];
```

$$\left[\{x = -1, y = 2\}, \left\{x = 2, y = \frac{1}{2}\right\}\right]$$

Now use the following command to substitute *each* of the solutions simultaneously into the original equations, eqns.

```
> map(subs, [soln], eqns);
```

$$[\{1 = 1, 3 = 3\}, \{1 = 1, 3 = 3\}]$$

This method can be valuable if your equation has many solutions, or if you are unsure of the number of solutions that a certain command will produce.

## Restricting Solutions

You can limit solutions by specifying inequalities with the `solve` command.

```
> solve({x^2=y^2},{x,y});
```

$$\{y = y, x = y\}, \{x = -y, y = y\}$$

```
> solve({x^2=y^2, x<>y},{x,y});
```

$$\{x = -y, \ y = y\}$$

Consider this system of five equations in five unknowns.

```
> eqn1 := x+2*y+3*z+4*t+5*u=41:
> eqn2 := 5*x+5*y+4*z+3*t+2*u=20:
> eqn3 := 3*y+4*z-8*t+2*u=125:
> eqn4 := x+y+z+t+u=9:
> eqn5 := 8*x+4*z+3*t+2*u=11:
```

Now solve the system for all variables.

```
> s1 := solve({eqn1,eqn2,eqn3,eqn4,eqn5}, {x,y,z,t,u});
```

$$s1 := \{x = 2, \ y = 3, \ u = 16, \ t = -11, \ z = -1\}$$

You can also choose to solve for a subset of the unknowns. Then Maple returns the solutions in terms of the other unknowns.

```
> s2 := solve({eqn1,eqn2,eqn3}, { x, y, z});
```

$$s2 := \{z = -7\,t - \frac{59}{13}\,u - \frac{70}{13}, \ y = 12\,t + \frac{70}{13}\,u + \frac{635}{13},$$

$$x = -7\,t - \frac{28}{13}\,u - \frac{527}{13}\}$$

## Exploring Solutions

You can explore the parametric solutions found at the end of the previous section. For example, evaluate the solution at $u = 1$ and $t = 1$.

```
> eval( s2, {u=1,t=1} );
```

$$\left\{x = \frac{-646}{13}, \ y = \frac{861}{13}, \ z = \frac{-220}{13}\right\}$$

As in *Verifying Solutions* on page 62, suppose that you require the solutions from solve in a particular order. Since you cannot fix the order of elements in a set, solve will not necessarily return your solutions in the order $x$, $y$, $z$. However, lists do preserve order; thus, try the following.

```
> eval( [x,y,z], s2 );
```

$$\left[-7\,t - \frac{28}{13}\,u - \frac{527}{13}, \ 12\,t + \frac{70}{13}\,u + \frac{635}{13}, \ -7\,t - \frac{59}{13}\,u - \frac{70}{13}\right]$$

This command not only fixed the order, but it also extracted the right-hand side of the equations (of course you still know which solution is for which variable because of their order). This capability is particularly useful should you wish to plot the solution surface.

```
> plot3d(%, u=0..2, t=0..2, axes=BOXED);
```

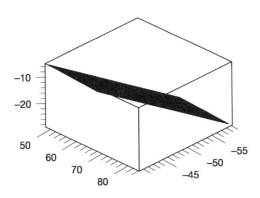

## The `unapply` Command

Suppose that you want to do even more exploring. For convenience, define $x = x(u, t)$, $y = y(u, t)$, and $z = z(u, t)$; that is, turn the solutions into functions. Recall that you may easily select a solution *expression* for a particular variable using `eval`.

```
> eval( x, s2 );
```

$$-7t - \frac{28}{13}u - \frac{527}{13}$$

However, this is an *expression* for $x$ and not a function.

```
> x(1,1);
```

$$x(1, 1)$$

To turn the expression into a function you need another very important command, `unapply`. To use it, provide `unapply` with the expression *and* the variables of which Maple should make it a function. Thus, for example

```
> f := unapply(x^2 + y^2 + 4, x, y);
```

$$f := (x, y) \rightarrow x^2 + y^2 + 4$$

produces the function, $f$, of $x$ and $y$, which maps $(x, y)$ to $x^2 + y^2 + 4$. This new function is easy to use.

```
> f(a,b);
```

$$a^2 + b^2 + 4$$

Thus, to make your solution for $x$ a function of both $u$ and $t$, the first step is to obtain the *expression* for $x$, as above.

```
> eval(x, s2);
```

$$-7t - \frac{28}{13}u - \frac{527}{13}$$

Then use `unapply` to turn it into a function of $u$ and $t$

```
> x := unapply(%, u, t);
```

$$x := (u, t) \rightarrow -7t - \frac{28}{13}u - \frac{527}{13}$$

```
> x(1,1);
```

$$\frac{-646}{13}$$

You can create the functions $y$ and $z$ in the same manner.

```
> eval(y,s2);
```

$$12t + \frac{70}{13}u + \frac{635}{13}$$

```
> y := unapply(%,u,t);
```

$$y := (u, t) \rightarrow 12t + \frac{70}{13}u + \frac{635}{13}$$

```
> eval(z,s2);
```

$$-7t - \frac{59}{13}u - \frac{70}{13}$$

```
> z := unapply(%, u, t);
```

$$z := (u, t) \rightarrow -7t - \frac{59}{13}u - \frac{70}{13}$$

```
> y(1,1), z(1,1);
```

$$\frac{861}{13}, \frac{-220}{13}$$

## The `assign` Command

The `assign` command also allocates values to unknowns. For example, instead of defining $x$, $y$, and $z$ as functions, assign each to the expression on the right-hand side of the corresponding equation.

```
> assign( s2 );
```

```
> x, y, z;
```

$$-7t - \frac{28}{13}u - \frac{527}{13}, \quad 12t + \frac{70}{13}u + \frac{635}{13}, \quad -7t - \frac{59}{13}u - \frac{70}{13}$$

Think of the assign command as turning the "=" signs in the solution set into ":=" signs.

assign is convenient if you wish to assign names to expressions. *Remember, though, that while this command is useful for quickly assigning solutions, it cannot create functions.*

This next example incorporates solving differential equations, which *Differential Equations:* dsolve on page 82 discusses in further detail. To begin, assign the solution.

```
> s3 := dsolve( {diff(f(x),x)=6*x^2+1, f(0)=0}, {f(x)} );
```

$$s3 := f(x) = 2x^3 + x$$

```
> assign( s3 );
```

However, you have yet to create a function.

```
> f(x);
```

$$2x^3 + x$$

produces the expected answer, but despite appearances, f(x) is simply a name for the *expression* $2x^3 + x$ and *not* a *function*. Call the function $f$ using an argument other than $x$.

```
> f(1);
```

$$f(1)$$

The reason for this apparently odd behavior is that assign asks Maple to do the assignment

```
> f(x) := 2*x^3 + x;
```

$$f(x) := 2x^3 + x$$

which is not at all the same as the assignment

```
> f := x -> 2*x^3 + x;
```

$$f := x \to 2x^3 + x$$

The former defines the value of the function $f$ for only the special argument $x$. The latter defines the function $f : x \mapsto 2x^3 + x$ so that it works whether you say $f(x)$, $f(y)$, or $f(1)$.

To define the solution $f$ as a function of $x$ use unapply.

```
> eval(f(x),s3);
```

$$2x^3 + x$$

```
> f := unapply(%, x);
```

$$f := x \rightarrow 2x^3 + x$$

```
> f(1);
```

$$3$$

## The RootOf Command

Maple occasionally returns solutions in terms of the RootOf command. The following example demonstrates this point.

```
> solve({x^5 - 2*x + 3 = 0},{x});
```

$$\{x = \text{RootOf}(\_Z^5 - 2\_Z + 3)\}$$

RootOf(*expr*) is a placeholder for all the roots of *expr*. What Maple is telling you is that $x$ is a root of the polynomial $z^5 - 2z + 3$. This can be useful if you are doing your algebra over a field other than the complex numbers. For now, focus on the complex roots by using the allvalues and evalf commands to obtain an explicit form of the solution.

```
> allvalues(%);
```

$$\{x = -1.423605849\},$$
$$\{x = -.2467292569 - 1.320816347\,I\},$$
$$\{x = -.2467292569 + 1.320816347\,I\},$$
$$\{x = .9585321812 - .4984277790\,I\},$$
$$\{x = .9585321812 + .4984277790\,I\}$$

```
> evalf(%);
```

$$\{x = -1.423605849\},$$
$$\{x = -.2467292569 - 1.320816347\,I\},$$
$$\{x = -.2467292569 + 1.320816347\,I\},$$
$$\{x = .9585321812 - .4984277790\,I\},$$
$$\{x = .9585321812 + .4984277790\,I\}$$

A general solution for the roots of degree five polynomials does not exist. Consequently, Maple can return only floating-point estimates.

## 3.2  Solving Numerically: `fsolve`

`fsolve` is the numeric equivalent of `solve`. `fsolve` calculates the equation(s) using a variation on Newton's method, producing approximate (floating-point) solutions.

```
> fsolve({cos(x)-x = 0}, {x});
```

$$\{x = .7390851332\}$$

For a general equation, `fsolve` searches for a single real root. For a polynomial, however, it looks for all *real* roots.

```
> poly :=3*x^4 - 16*x^3 - 3*x^2 + 13*x + 16;
```

$$poly := 3\,x^4 - 16\,x^3 - 3\,x^2 + 13\,x + 16$$

```
> fsolve({poly},{x});
```

$$\{x = 1.324717957\}, \{x = 5.333333333\}$$

To look for more than one root of a general equation, try the following approach: divide the original equation by the root, and solve again.

```
> fsolve({sin(x)=0}, {x});
```

$$\{x = 0\}$$

```
> x1 := eval(x, %);
```

$$x1 := 0$$

```
> fsolve({sin(x)/(x-x1)=0}, {x});
```

$$\{x = -3.141592654\}$$

```
> x2 := eval(x, %);
```

$$x2 := -3.141592654$$

```
> fsolve({sin(x)/(x-x1)/(x-x2)=0}, {x});
```

$$\{x = 3.141592654\}$$

You can restrict Maple to look for only a certain number of roots in a polynomial by setting the option `maxsols`.

```
> fsolve({poly}, {x}, maxsols=1);
```

$$\{x = 1.324717957\}$$

The option `complex` forces Maple to search for complex roots in addition to real roots.

```
> fsolve({poly}, {x}, complex);
```

$${x = -.6623589786 - .5622795121 \, I},$$

$${x = -.6623589786 + .5622795121 \, I},$$

$${x = 1.324717957}, {x = 5.333333333}$$

You can also specify a range in which to look for a root.

```
> fsolve({cos(x)=0}, {x}, Pi..2*Pi);
```

$${x = 4.712388980}$$

In some cases, `fsolve` may fail to find a root even if one exists. In these cases, specifying a range should help. To increase the accuracy of the solutions, you can increase the value of the special variable, `Digits`. Note that in the following example the solution is not guaranteed to be accurate to thirty digits, but rather, Maple performs all steps in the solution to at least thirty significant digits rather than the default of ten.

```
> Digits := 30;
```

$$Digits := 30$$

```
> fsolve({cos(x)=0}, {x});
```

$${x = 1.57079632679489661923132169164}$$

## Limitations on `solve`

`solve` cannot handle every problem. Remember that Maple's approach is algorithmic, and it does not necessarily have the ability to use the "tricks" that you might use when solving the problem by hand.

Mathematically, polynomials of degree five or higher do not have a general solution. Maple does its best to solve them anyway, but you may have to resort to a numerical solution.

Solving trigonometric equations can also be difficult, and Maple does have some limitations. In fact, working with all transcendental equations is quite difficult.

```
> solve({sin(x)=0}, {x});
```

$${x = 0}$$

Note that Maple returns only one of an infinite number of solutions. However, with the `fsolve` command you can specify the range in which to look for a solution. Thereby you may gain more control over the solution.

```
> fsolve({sin(x)=0}, {x}, 3..4);
```

$$\{x = 3.141592653589793238462643383328\}$$

If Maple cannot find a solution, it returns nothing. This does not mean that a solution does not exist, but that Maple could not find one. In the following example, a solution clearly exists, but Maple could not find it.

```
> solve({-x^5=-1+sin(x)}, {x});
> plot({-x^5, -1+sin(x)}, x=-1..1);
```

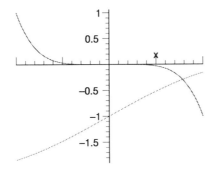

However, `fsolve` comes up with the solution.

```
> fsolve({-x^5=-1+sin(x)}, {x});
```

$$\{x = .7831075605309905337614360947 22\}$$

These types of problems are common to all symbolic computation systems, and are symptoms of the natural limitations of an algorithmic approach to equation solving.

When using `solve`, remember to check your results. The next example highlights a misunderstanding that can arise because of Maple's handling of removable singularities.

```
> expr := (x-1)^2/(x^2-1);
```

$$expr := \frac{(x-1)^2}{x^2 - 1}$$

Maple finds a solution

```
> soln := solve({expr=0},{x});
```

$$soln := \{x = 1\}$$

but when you evaluate the expression at 1, you get 0/0.

```
> eval(expr, soln);
```

Error, division by zero

The limit shows that $x = 1$ is nearly a solution.

```
> Limit(expr, x=1);
```

$$\lim_{x \to 1} \frac{(x - 1)^2}{x^2 - 1}$$

```
> value (%);
```

$$0$$

Even the plot has problems, unless you specify discont=true.

```
> plot(expr, x=-5..5, y=-10..10);
```

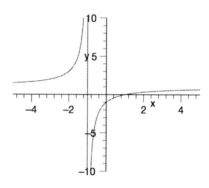

Maple removed the singularity $x = 1$ from the expression before solving it. No matter what method or tools you use to solve equations, always check your results. Fortunately these checks are easy to do in Maple.

## 3.3 Other Solvers

Maple contains a number of specialized solve commands. Since you are not as likely to find these as useful as the more general commands, solve and fsolve, this section only briefly mentions some of them. If you require more details on any of these commands, take advantage of the on-line help by entering ? and the command name at the Maple prompt.

### Finding Integer Solutions

The isolve command finds integer solutions to equations, solving for all unknowns in the expression(s).

```
> isolve({3*x-4*y=7});
```

$$\{x = 5 + 4\_N1, \ y = 2 + 3\_N1\}$$

Maple uses the global names _N1, ..., _Nn to denote parameters of the solution.

## Finding Solutions Modulo $m$

The msolve command solves equations in the integers modulo $m$ (the positive representation for integers), solving for all unknowns in the expression(s).

```
> msolve({3*x-4*y=1,7*x+y=2},17);
```

$$\{y = 6, \ x = 14\}$$

Maple uses global names _NN1, ... _NNn for parameters in the solutions.

```
> msolve({2^n=3},19);
```

$$\{n = 13 + 18\_NN1\tilde{}\}$$

The tilde (˜) on _NN1 indicates that msolve has placed an assumption on _NN1; in this case that _NN1 is an integer.

```
> about( _NN1 );

Originally _NN1, renamed _NN1~:
   is assumed to be: integer
```

*The Assume Facility* on page 166 describes how you can place assumptions on unknowns yourself.

## Solving Recurrence Relations

The rsolve command solves recurrence equations, returning an expression for the general term of the function.

```
> rsolve({f(n)=f(n-1)+f(n-2),f(0)=1,f(1)=1},{f(n)});
```

$$\left\{ f(n) = -\frac{2}{5} \frac{\sqrt{5}\left(-\dfrac{2}{1-\sqrt{5}}\right)^n}{1-\sqrt{5}} + \frac{2}{5} \frac{\sqrt{5}\left(-\dfrac{2}{1+\sqrt{5}}\right)^n}{1+\sqrt{5}} \right\}$$

See also ?LREtools.

## 3.4  Polynomials

A *polynomial* in Maple is an expression containing unknowns. Each term in the polynomial contains a product of the unknowns. For example, should the polynomial contain only one unknown, $x$, then the terms might contain $x^3$, $x^1 = x$, and $x^0 = 1$ as in the case of the polynomial $x^3 - 2x + 1$. If more than one unknown exists, then a term may also contain a product of the unknowns, as in the polynomial $x^3 + 3x^2y + y^2$. Coefficients can be integers (as in the examples above), rational numbers, irrational numbers, floating-point numbers, complex numbers, or even other variables.

```
> x^2 - 1;
```

$$x^2 - 1$$

```
> x + y + z;
```

$$x + y + z$$

```
> 1/2*x^2 - sqrt(3)*x - 1/3;
```

$$\frac{1}{2}x^2 - \sqrt{3}\,x - \frac{1}{3}$$

```
> (1 - I)*x + 3 + 4*I;
```

$$(1 - I)\,x + 3 + 4\,I$$

```
> a*x^4 + b*x^3 + c*x^2 + d*x + f;
```

$$a\,x^4 + b\,x^3 + c\,x^2 + d\,x + f$$

Maple possesses commands for many kinds of manipulations and mathematical calculations with polynomials; the following sections investigate some of them.

### Sorting and Collecting

The `sort` command arranges a polynomial into descending order of powers of the unknowns. Rather than making another copy of the polynomial with the terms in order, `sort` modifies the way Maple stores the original polynomial in memory. In other words, if you display your polynomial after sorting it, you will find that it retains the new order.

```
> sort_poly := x + x^2 - x^3 + 1 - x^4;
```

$$sort\_poly := x + x^2 - x^3 + 1 - x^4$$

```
> sort(sort_poly);
```

$$-x^4 - x^3 + x^2 + x + 1$$

```
> sort_poly;
```

$$-x^4 - x^3 + x^2 + x + 1$$

Maple sorts multivariate polynomials in two ways. The default method sorts them by total degree of the terms. Thus, $x^2 y^2$ will come before both $x^3$ and $y^3$. The other option sorts by pure lexicographic order (plex). When you choose this option, the sort deals first with the powers of the first variable in the variable list (second argument) and then with the powers of the second variable. The difference between these sorts is best shown by an example.

```
> mul_var_poly := y^3 + x^2*y^2 + x^3;
```

$$mul\_var\_poly := y^3 + x^2 y^2 + x^3$$

```
> sort(mul_var_poly, [x,y]);
```

$$x^2 y^2 + x^3 + y^3$$

```
> sort(mul_var_poly, [x,y], 'plex');
```

$$x^3 + x^2 y^2 + y^3$$

collect brings together coefficients of like powers in a polynomial. For example should the terms $ax$ and $bx$ both be in the polynomial, Maple collects them together into $(a + b)x$.

```
> big_poly:=x*y + z*x*y + y*x^2 - z*y*x^2 + x + z*x;
```

$$big\_poly := x y + z x y + y x^2 - z y x^2 + x + z x$$

```
> collect(big_poly, x);
```

$$(y - z y) x^2 + (y + z y + 1 + z) x$$

```
> collect(big_poly, z);
```

$$(x y - y x^2 + x) z + x y + y x^2 + x$$

## Mathematical Operations

You can perform many mathematical operations on polynomials. Among the most fundamental is division; that is, to divide one polynomial into another and determine the quotient and remainder. Maple provides the commands rem and quo to find the remainder and quotient of a polynomial division.

```
> r := rem(x^3+x+1, x^2+x+1, x);
```

$$r := 2 + x$$

```
> q := quo(x^3+x+1, x^2+x+1, x);
```

$$q := x - 1$$

```
> collect( (x^2+x+1) * q + r, x );
```

$$x^3 + x + 1$$

On the other hand, sometimes all you need to know is whether or not one polynomial divides into another polynomial exactly. divide tests for exact polynomial division.

```
> divide(x^3 - y^3, x - y);
```

$$true$$

```
> rem(x^3 - y^3, x - y, x);
```

$$0$$

You evaluate polynomials at values as you would with any expression, by using eval.

```
> poly := x^2 + 3*x - 4;
```

$$poly := x^2 + 3x - 4$$

```
> eval(poly, x=2);
```

$$6$$

```
> mul_var_poly := y^2*x - 2*y + x^2*y + 1;
```

$$mul\_var\_poly := y^2 x - 2y + y x^2 + 1$$

```
> eval(mul_var_poly, {y=1,x=-1});
```

$$-1$$

## Coefficients and Degrees

The commands degree and coeff determine the degree of a polynomial and provide a mechanism for extracting coefficients.

```
> poly := 3*z^3 - z^2 + 2*z - 3*z + 1;
```

$$poly := 3z^3 - z^2 - z + 1$$

```
> coeff(poly, z^2);
```

$$-1$$

**TABLE 3.1** Commands for Finding Polynomial Coefficients

| | |
|---|---|
| coeff | extract coefficient |
| lcoeff | find the leading coefficient |
| tcoeff | find the trailing coefficient |
| coeffs | return a sequence of all the coefficients |
| degree | determine the [highest] degree of the polynomial |
| ldegree | determine the lowest degree of the polynomial |

```
> degree(poly);
```

$$3$$

## Root Finding and Factorization

solve determines the roots of a polynomial; whereas, factor expresses the polynomial in fully factored form.

```
> poly1 := x^6 - x^5 - 9*x^4 + x^3 + 20*x^2 + 12*x;
```

$$poly1 := x^6 - x^5 - 9x^4 + x^3 + 20x^2 + 12x$$

```
> factor(poly1);
```

$$x(x-2)(x-3)(x+2)(x+1)^2$$

```
> poly2 := (x + 3);
```

$$poly2 := x + 3$$

```
> poly3 := expand(poly2^6);
```

$$poly3 :=$$
$$x^6 + 18x^5 + 135x^4 + 540x^3 + 1215x^2 + 1458x + 729$$

```
> factor(poly3);
```

$$(x+3)^6$$

```
> solve({poly3=0}, {x});
```

$$\{x = -3\}, \{x = -3\}, \{x = -3\}, \{x = -3\}, \{x = -3\}, \{x = -3\}$$

```
> factor(x^3 + y^3);
```

$$(x+y)(x^2 - xy + y^2)$$

Maple factors the polynomial over the ring implied by the coefficients (integers, rationals, etc.). The factor command also allows you to specify

**TABLE 3.2** Some Other Functions for Working with Polynomials

| | |
|---|---|
| content | content of a multivariate polynomial |
| compoly | polynomial decomposition |
| discrim | discriminant of a polynomial |
| gcd | greatest common divisor |
| gcdex | extended Euclidean algorithm |
| interp | polynomial interpolation |
| lcm | least common multiple |
| norm | norm of a polynomial |
| prem | pseudo-remainder |
| primpart | primitive part of a multivariate polynomial |
| randpoly | random polynomial |
| recipoly | reciprocal polynomial |
| resultant | resultant of two polynomials |
| roots | roots over an algebraic number field |
| sqrfree | square-free factorization |

an algebraic number field over which to factor the polynomial. See the help page (?factor) for more information.

## 3.5  Calculus

Maple provides many powerful tools for solving problems in calculus. For example, Maple is useful for computing limits of functions. Compute the limit of a rational function as $x$ approaches 1.

```
> f := x -> (x^2-2*x+1)/(x^4 + 3*x^3 - 7*x^2 + x+2);
```

$$f := x \rightarrow \frac{x^2 - 2x + 1}{x^4 + 3x^3 - 7x^2 + x + 2}$$

```
> Limit(f(x), x=1);
```

$$\lim_{x \to 1} \frac{x^2 - 2x + 1}{x^4 + 3x^3 - 7x^2 + x + 2}$$

```
> value(%);
```

$$\frac{1}{8}$$

Taking the limit of an expression from either the positive or the negative direction is also possible. For example, consider the limit of $\tan(x)$ as $x$ approaches $\pi/2$.

Calculate the left-hand limit using the option left.

```
> Limit(tan(x), x=Pi/2, left);
```

$$\lim_{x \to (1/2\,\pi)-} \tan(x)$$

```
> value(%);
```
$$\infty$$

Do the same for the right-hand limit.

```
> Limit(tan(x), x=Pi/2, right);
```
$$\lim_{x \to (1/2\,\pi)+} \tan(x)$$

```
> value(%);
```
$$-\infty$$

Another operation easily performed in Maple is the creation of series approximations of a function. For example, use the function

```
> f := x -> sin(4*x)*cos(x);
```
$$f := x \to \sin(4\,x)\cos(x)$$

```
> fs1 := series(f(x), x=0);
```
$$fs1 := 4\,x - \frac{38}{3}\,x^3 + \frac{421}{30}\,x^5 + O(x^6)$$

Note that, by default, the `series` command generates an order 6 polynomial. By changing the value of the special variable, `Order`, you can increase or decrease the order of a polynomial series.

Using `convert(fs1, polynom)` removes the order term from the series so that Maple can plot it.

```
> p := convert(fs1,polynom);
```
$$p := 4\,x - \frac{38}{3}\,x^3 + \frac{421}{30}\,x^5$$

```
> plot({f(x), p},x=-1..1, -2..2,
>       title="sin(4x) cos(x) vs. Series");
```

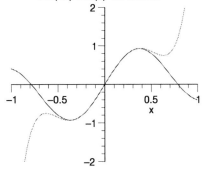

sin(4x) cos(x) vs. Series

If you increase the order of truncation of the series to 12 and try again, you see the expected improvement in the accuracy of the approximation.

```
> Order := 12;
```

$$Order := 12$$

```
> fs1 := series(f(x), x=0);
```

$$fs1 := 4x - \frac{38}{3}x^3 + \frac{421}{30}x^5 - \frac{10039}{1260}x^7 + \frac{246601}{90720}x^9 -$$
$$\frac{6125659}{9979200}x^{11} + O(x^{12})$$

```
> p := convert(fs1,polynom);
```

$$p := 4x - \frac{38}{3}x^3 + \frac{421}{30}x^5 - \frac{10039}{1260}x^7 + \frac{246601}{90720}x^9$$
$$- \frac{6125659}{9979200}x^{11}$$

```
> plot({f(x), p}, x=-1..1, -2..2,
>     title="sin(4x) cos(x) vs. Series of Order 12");
```

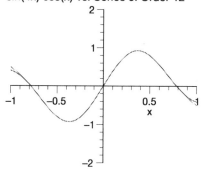

sin(4x) cos(x) vs. Series of Order 12

Maple can symbolically compute derivatives and integrals. For example, differentiate an expression, integrate its result, and compare it with the original expression.

```
> f := x -> x*sin(a*x) + b*x^2;
```

$$f := x \rightarrow x \sin(a\,x) + b\,x^2$$

```
> Diff(f(x),x);
```

$$\frac{\partial}{\partial x}(x \sin(a\,x) + b\,x^2)$$

```
> df := value(%);
```

$$df := \sin(a\,x) + x\cos(a\,x)\,a + 2\,b\,x$$

```
> Int(df, x);
```

$$\int \sin(a\,x) + x\cos(a\,x)\,a + 2\,b\,x\,dx$$

```
> value(%);
```

$$-\frac{\cos(a\,x)}{a} + \frac{\cos(a\,x) + a\,x\sin(a\,x)}{a} + b\,x^2$$

```
> simplify(%);
```

$$x\sin(a\,x) + b\,x^2$$

You can also perform definite integrations. For example, recompute the previous integral within the interval $x = 1$ to $x = 2$.

```
> Int(df,x=1..2);
```

$$\int_1^2 \sin(a\,x) + x\cos(a\,x)\,a + 2\,b\,x\,dx$$

```
> value(%);
```

$$2\sin(2\,a) + 3\,b - \sin(a)$$

Consider a more complicated integral.

```
> Int(exp(-x^2), x);
```

$$\int e^{(-x^2)}\,dx$$

```
> value(%);
```

$$\frac{1}{2}\sqrt{\pi}\,\mathrm{erf}(x)$$

Sometimes Maple is not sure whether a variable is real or complex and thus returns an unexpected result.

```
> g := t -> exp(-a*t)*ln(t);
```

$$g := t \rightarrow e^{(-a\,t)}\ln(t)$$

```
> Int (g(t), t=0..infinity);
```

$$\int_0^\infty e^{(-a\,t)}\ln(t)\,dt$$

```
> value(%);
```

$$\lim_{t \to \infty} -\frac{e^{(-at)} \ln(t) + \text{Ei}(1,\, at) + \gamma + \ln(a)}{a}$$

Here Maple assumes that the parameter a is a complex number; hence the complicated answer and warning messages. For situations where you know that a is a positive, real number, tell Maple by using the assume command.

```
> assume(a > 0):
> ans := Int(g(t), t=0..infinity);
```

$$ans := \int_0^\infty e^{(-\tilde{a}\, t)} \ln(t)\, dt$$

```
> value(%);
```

$$-\frac{\ln(\tilde{a})}{\tilde{a}} - \frac{\gamma}{\tilde{a}}$$

The result is much simpler. The only non-elementary term is the constant gamma. The tilde (˜) implies that a carries an assumption. Now remove the assumption in order to proceed to more examples. You must do this in two steps. The answer, ans, contains the a with assumptions. If you wish to reset and continue to use ans, then you must replace all occurrences of $\tilde{a}$ with $a$.

```
> ans := subs(a ='a', ans );
```

$$ans := \int_0^\infty e^{(-at)} \ln(t)\, dt$$

The first argument, a = 'a', deserves special attention. If you type a after making an assumption about a, Maple automatically assumes you mean $\tilde{a}$. In Maple, single quotes *delay evaluation*. In this case, they ensure that Maple interprets the second a as $a$ and not as $\tilde{a}$.

Now that you have removed the assumption on a inside ans, you can remove the assumption on a itself by assigning it to its own name.

```
> a := 'a':
```

Use single quotes here for the same reason as before. See also *The Assume Facility* on page 166.

## 3.6 Differential Equations: dsolve

Maple can symbolically solve many ordinary differential equations (ODEs), including initial value and boundary value problems.

Define an ODE.

```
> ode1 := {diff(y(t),t,t) + 5*diff(y(t),t) + 6*y(t)  = 0};
```

$$ode1 := \left\{ \left( \frac{\partial^2}{\partial t^2} y(t) \right) + 5 \left( \frac{\partial}{\partial t} y(t) \right) + 6\, y(t) = 0 \right\}$$

Define initial conditions.

```
> ic := {y(0)=0, D(y)(0)=1};
```

$$ic := \{ D(y)(0) = 1,\ y(0) = 0 \}$$

Solve with `dsolve`, using the `union` operator to form the union of the two sets.

```
> soln := dsolve(ode1 union ic, {y(t)});
```

$$soln := y(t) = -e^{(-3t)} + e^{(-2t)}$$

Should you wish to make use of this solution by evaluating it at points or by plotting it, remember to use the `unapply` command to define a proper Maple function. *Simple* `solve` on page 60 discusses this further.

You can conveniently extract a value from a solution set with the aid of `eval`.

```
> eval( y(t), soln );
```

$$-e^{(-3t)} + e^{(-2t)}$$

Now, use this fact to define $y$ as a function of $t$ using `unapply`:

```
> y1:= unapply(%, t );
```

$$y1 := t \rightarrow -e^{(-3t)} + e^{(-2t)}$$

```
> y1(a);
```

$$-e^{(-3a)} + e^{(-2a)}$$

Now verify that `y1` is indeed a solution to the ODE:

```
> eval(ode1, y=y1);
```

$$\{ 0 = 0 \}$$

and that `y1` satisfies the initial conditions.

```
> eval(ic, y=y1);
```

$$\{ 0 = 0,\ 1 = 1 \}$$

Another method for solution checking is also available, but it may seem puzzling at first. Assign `y` as the new solution name, instead of `y1`.

```
> y := unapply( eval(y(t), soln), t );
```

$$y := t \to -e^{(-3t)} + e^{(-2t)}$$

Now when you enter an equation containing y, Maple uses this function and evaluates the result, all in one step.

```
> ode1;
```

$$\{0 = 0\}$$

```
> ic;
```

$$\{0 = 0, \ 1 = 1\}$$

Should you wish to change the differential equation and try again, or should you no longer want this definition of $y(x)$, then you may remove the definition with the following command.

```
> y := 'y';
```

$$y := y$$

Maple also understands special functions, such as the Dirac delta or impulse function, used in physics.

```
> ode2 := 10^6*diff(y(x),x,x,x,x) = Dirac(x-2) -
>    Dirac(x-4);
```

$$ode2 := 1000000 \left( \frac{\partial^4}{\partial x^4} y(x) \right) = \text{Dirac}(x - 2) - \text{Dirac}(x - 4)$$

Specify boundary conditions

```
> bc := {y(0)=0, D(D(y))(0)=0, y(5)=0};
```

$$bc := \{y(0) = 0, \ y(5) = 0, \ (D^{(2)})(y)(0) = 0\}$$

and an initial value.

```
> iv := {D(D(y))(5)=0};
```

$$iv := \{(D^{(2)})(y)(5) = 0\}$$

```
> soln := dsolve({ode2} union bc union iv, {y(x)});
```

$$soln := y(x) = \frac{1}{6000000} \text{Heaviside}(x - 2) x^3$$

$$- \frac{1}{750000} \text{Heaviside}(x - 2) + \frac{1}{500000} \text{Heaviside}(x - 2) x$$

$$-\frac{1}{1000000}\,\text{Heaviside}(x-2)\,x^2$$

$$-\frac{1}{6000000}\,\text{Heaviside}(x-4)\,x^3$$

$$+\frac{1}{93750}\,\text{Heaviside}(x-4)-\frac{1}{125000}\,\text{Heaviside}(x-4)\,x$$

$$+\frac{1}{500000}\,\text{Heaviside}(x-4)\,x^2-\frac{1}{15000000}\,x^3$$

$$+\frac{1}{1250000}\,x$$

```
> eval(y(x), soln);
```

$$\frac{1}{6000000}\,\text{Heaviside}(x-2)\,x^3-\frac{1}{750000}\,\text{Heaviside}(x-2)$$

$$+\frac{1}{500000}\,\text{Heaviside}(x-2)\,x$$

$$-\frac{1}{1000000}\,\text{Heaviside}(x-2)\,x^2$$

$$-\frac{1}{6000000}\,\text{Heaviside}(x-4)\,x^3$$

$$+\frac{1}{93750}\,\text{Heaviside}(x-4)-\frac{1}{125000}\,\text{Heaviside}(x-4)\,x$$

$$+\frac{1}{500000}\,\text{Heaviside}(x-4)\,x^2-\frac{1}{15000000}\,x^3$$

$$+\frac{1}{1250000}\,x$$

```
> y := unapply(%, x);
```

$$y := x \rightarrow \frac{1}{6000000}\,\text{Heaviside}(x-2)\,x^3$$

$$-\frac{1}{750000}\,\text{Heaviside}(x-2)+\frac{1}{500000}\,\text{Heaviside}(x-2)\,x$$

$$-\frac{1}{1000000}\,\text{Heaviside}(x-2)\,x^2$$

$$-\frac{1}{6000000}\,\text{Heaviside}(x-4)\,x^3$$

$$+\frac{1}{93750}\,\text{Heaviside}(x-4)-\frac{1}{125000}\,\text{Heaviside}(x-4)\,x$$

$$+ \frac{1}{500000} \, \text{Heaviside}(x - 4) \, x^2 - \frac{1}{15000000} \, x^3$$

$$+ \frac{1}{1250000} \, x$$

This value of y satisfies the differential equation, the boundary conditions, and the initial value.

```
> ode2;
```

$$8 \, \text{Dirac}(2, \, x - 2) + 2 \, \text{Dirac}(2, \, x - 2) \, x^2$$

$$+ 16 \, \text{Dirac}(2, \, x - 4) \, x - 2 \, \text{Dirac}(2, \, x - 4) \, x^2$$

$$- \text{Dirac}(3, \, x - 2) \, x^2 - \frac{1}{6} \, \text{Dirac}(3, \, x - 4) \, x^3$$

$$+ \frac{32}{3} \, \text{Dirac}(3, \, x - 4) - \frac{4}{3} \, \text{Dirac}(3, \, x - 2)$$

$$- 32 \, \text{Dirac}(2, \, x - 4) + 4 \, \text{Dirac}(x - 2) - 4 \, \text{Dirac}(x - 4)$$

$$- 8 \, \text{Dirac}(2, \, x - 2) \, x - 8 \, \text{Dirac}(3, \, x - 4) \, x$$

$$+ 2 \, \text{Dirac}(3, \, x - 4) \, x^2 + \frac{1}{6} \, \text{Dirac}(3, \, x - 2) \, x^3$$

$$+ 2 \, \text{Dirac}(3, \, x - 2) \, x + 6 \, \text{Dirac}(1, \, x - 2) \, x$$

$$- 6 \, \text{Dirac}(1, \, x - 4) \, x - 12 \, \text{Dirac}(1, \, x - 2)$$

$$+ 24 \, \text{Dirac}(1, \, x - 4) = \text{Dirac}(x - 2) - \text{Dirac}(x - 4)$$

```
> simplify(%);
```

$$\text{Dirac}(x - 2) - \text{Dirac}(x - 4) = \text{Dirac}(x - 2) - \text{Dirac}(x - 4)$$

```
> bc;
```

$$\{0 = 0\}$$

```
> iv;
```

$$\{0 = 0\}$$

```
> plot(y(x), x=0..5, axes=BOXED,
```

> `title="Solution to a 4th Order BVP");`

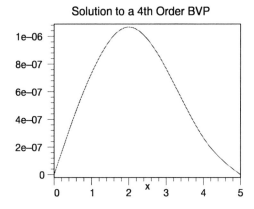

Solution to a 4th Order BVP

You should unassign y now since you are done with it.

> `y := 'y';`

$$y := y$$

Maple can also solve systems of differential equations. For example, solve the following system of two simultaneous, second order equations.

> `de_sys := { diff(y(x),x,x)=z(x), diff(z(x),x,x)=y(x) };`

$$de\_sys := \left\{ \frac{\partial^2}{\partial x^2}\, y(x) = z(x),\ \frac{\partial^2}{\partial x^2}\, z(x) = y(x) \right\}$$

> `soln := dsolve(de_sys, {z(x),y(x)});`

$$soln := \{ z(x) = \frac{1}{4}\, \_C1\, e^{(-x)} + \frac{1}{4}\, \_C1\, e^x - \frac{1}{2}\, \_C1 \cos(x)$$

$$-\frac{1}{2}\, \_C2 \sin(x) + \frac{1}{4}\, \_C2\, e^x - \frac{1}{4}\, \_C2\, e^{(-x)} + \frac{1}{4}\, \_C3\, e^{(-x)}$$

$$+\frac{1}{4}\, \_C3\, e^x + \frac{1}{2}\, \_C3 \cos(x) - \frac{1}{4}\, \_C4\, e^{(-x)} + \frac{1}{4}\, \_C4\, e^x$$

$$+\frac{1}{2}\, \_C4 \sin(x),\ y(x) = \frac{1}{4}\, \_C1\, e^{(-x)} + \frac{1}{4}\, \_C1\, e^x$$

$$+\frac{1}{2}\, \_C1 \cos(x) - \frac{1}{4}\, \_C2\, e^{(-x)} + \frac{1}{4}\, \_C2\, e^x + \frac{1}{2}\, \_C2 \sin(x)$$

$$+\frac{1}{4}\, \_C3\, e^{(-x)} + \frac{1}{4}\, \_C3\, e^x - \frac{1}{2}\, \_C3 \cos(x) - \frac{1}{2}\, \_C4 \sin(x)$$

$$+\frac{1}{4}\, \_C4\, e^x - \frac{1}{4}\, \_C4\, e^{(-x)} \}$$

If you solve the system without providing additional conditions, Maple automatically generates the appropriate constants _C1, ..., _C4.

Again, observe that you can easily extract and define the solutions with the aid of eval and unapply:

```
> y := unapply(eval(y(x), soln), x );
```

$$y := x \rightarrow \frac{1}{4} \_C1 \, e^{(-x)} + \frac{1}{4} \_C1 \, e^x + \frac{1}{2} \_C1 \cos(x)$$

$$- \frac{1}{4} \_C2 \, e^{(-x)} + \frac{1}{4} \_C2 \, e^x + \frac{1}{2} \_C2 \sin(x) + \frac{1}{4} \_C3 \, e^{(-x)}$$

$$+ \frac{1}{4} \_C3 \, e^x - \frac{1}{2} \_C3 \cos(x) - \frac{1}{2} \_C4 \sin(x) + \frac{1}{4} \_C4 \, e^x$$

$$- \frac{1}{4} \_C4 \, e^{(-x)}$$

```
> y(1);
```

$$\frac{1}{4} \_C1 \, e^{(-1)} + \frac{1}{4} \_C1 \, e + \frac{1}{2} \_C1 \cos(1) - \frac{1}{4} \_C2 \, e^{(-1)}$$

$$+ \frac{1}{4} \_C2 \, e + \frac{1}{2} \_C2 \sin(1) + \frac{1}{4} \_C3 \, e^{(-1)} + \frac{1}{4} \_C3 \, e$$

$$- \frac{1}{2} \_C3 \cos(1) - \frac{1}{2} \_C4 \sin(1) + \frac{1}{4} \_C4 \, e - \frac{1}{4} \_C4 \, e^{(-1)}$$

and you can undefine it again when you are finished with it.

```
> y := 'y';
```

$$y := y$$

## 3.7 The Organization of Maple

When you start Maple, it loads only the *kernel*. The kernel is the base of Maple's system. It contains fundamental and primitive commands: the Maple language interpreter (which converts the commands you type into machine instructions your computer processor can understand), algorithms for numerical calculation, and routines to display results and perform other input and output operations.

The kernel consists of highly optimized C code—approximating 10% of the system's total size. Maple programmers have deliberately kept the size of the kernel small for speed and efficiency. The Maple kernel implements the most frequently used routines for integer and rational arithmetic and simple polynomial calculations.

The remaining 90% of Maple's mathematical knowledge is written in the Maple language and resides in the Maple library. Maple's library divides into three groups: the *main* library, the *miscellaneous* library, and the packages. These groups of functions sit above the kernel.

The *main* library contains the most frequently used Maple commands (other than those in the kernel). These commands load upon demand— you do not need to explicitly load them. The Maple language produces very compact procedures that read with no observable delay, so you are not likely to notice which commands are C-coded kernel commands and which are loaded from the library.

The *miscellaneous* library consists of many less frequently used mathematical commands. Since they are not `readlib`-defined, you must explicitly load them. Use the `readlib` command with the following syntax.

> `readlib(`*cmd*`)`

Here *cmd* is the command you want Maple to load from the library.

The last commands in the library are in the packages. Each one of Maple's numerous packages contains a group of commands for related calculations. For example, the `linalg` package contains commands for the manipulation of matrices.

You can use a command from a package in three ways.

1. Use the complete name of the package and the desired command.

> *package*`[`*cmd*`](` ... `)`

2. Activate the short names for all the commands in a package using the `with` command.

> `with(`*package*`)`

Then use the short name for the command.

> *cmd*`(...)`

3. Activate the short name for a single command from a package.

> `with(`*package*`, `*cmd*`)`

Then use the short form of the command name.

> *cmd*`(...)`

This next example uses the `distance` command in the `student` package to calculate the distance between two points.

```
> with(student);
```

$$[D, \textit{Diff}, \textit{Doubleint}, \textit{Int}, \textit{Limit}, \textit{Lineint}, \textit{Product}, \textit{Sum},$$

$$\textit{Tripleint}, \textit{changevar}, \textit{combine}, \textit{completesquare},$$

$$\textit{distance}, \textit{equate}, \textit{extrema}, \textit{integrand}, \textit{intercept}, \textit{intparts},$$

$$\textit{isolate}, \textit{leftbox}, \textit{leftsum}, \textit{makeproc}, \textit{maximize},$$

$$\textit{middlebox}, \textit{middlesum}, \textit{midpoint}, \textit{minimize}, \textit{powsubs},$$

$$\textit{rightbox}, \textit{rightsum}, \textit{showtangent}, \textit{simpson}, \textit{slope},$$

$$\textit{summand}, \textit{trapezoid}, \textit{value}]$$

```
> distance([1,1],[3,4]);
```

$$\sqrt{13}$$

The long form of package command names are initially `readlib`-defined; Maple loads the individual commands only when you explicitly need them. When you load a package, Maple `readlib`-defines all the short names of commands in the package.

When you load a package using `with(`*package*`)`, you see a list of all the short names of the commands in the package. Plus, Maple warns you if it has redefined any pre-existing names.

## 3.8  The Maple Packages

Maple's built-in packages of specialized commands perform tasks from a huge variety of disciplines, from Student Calculus to General Relativity Theory. The examples in this section are not intended to be comprehensive. They are simply examples of a few commands in selected packages, to give you a taste of the things Maple can do.

### List of Packages

The following list of packages can also be found by reading the help page for `?packages`. For a full list of commands in a particular package, see the help page, `?`*packagename*.

`algcurves`   algebraic curves tools for studying the one-dimensional algebraic varieties (curves) defined by multi-variate polynomials.

`codegen`   tools for creating, manipulating, and translating Maple procedures into other languages. Includes automatic differentiation, code optimization, translation into C and Fortran, etc.

combinat  combinatorial functions, including commands for calculating permutations and combinations of lists, and partitions of integers. (Use combstruct package instead, where possible.)

combstruct  commands for generating and counting combinatorial structures, as well as determining generating function equations for such counting.

context  tools for building and modifying context-sensitive menus in Maple's graphical user interface (e.g., when right-clicking on an output expression).

DEtools  tools for manipulating, solving, and plotting systems of differential equations; phase portraits and field plots.

difforms  commands for handling differential forms; for problems in differential geometry.

Domains  commands to create *domains of computation*; supports computing with polynomials, matrices, and series over number rings, finite fields, polynomial rings, and matrix rings.

finance  commands for financial computations.

GaussInt  commands for working with Gaussian Integers; that is, numbers of the form $a + bI$ where $a$ and $b$ are integers. Commands for finding GCDs, factoring, and primality testing.

genfunc  commands for manipulating rational generating functions.

geom3d  commands for three-dimensional Euclidean geometry; to define and manipulate points, lines, planes, triangles, spheres, polyhedra, etc. in three dimensions.

geometry  commands for two-dimensional Euclidean geometry; to define and manipulate points, lines, triangles, and circles in two dimensions.

Groebner  commands for Gröbner basis computations; in particular tools for Ore algebras and D-modules.

group  commands for working with permutation groups and finitely-presented groups.

inttrans  commands for working with integral transforms and their inverses.

liesymm  commands for characterizing the contact symmetries of systems of partial differential equations.

linalg  over 100 commands for matrix and vector manipulation; everything from adding two matrices to symbolic eigenvectors and eigenvalues.

`logic` commands for constructing and working with Boolean expressions and functions.

`LREtools` commands for manipulating, plotting, and solving linear recurrence equations.

`Matlab` commands to use several of Matlab's numerical matrix functions, including eigenvalues and eigenvectors, determinants, and LU-decomposition. (Only accessible if Matlab is installed on your system.)

`networks` tools for constructing, drawing, and analyzing combinatorial networks. Facilities for handling directed graphs, and arbitrary expressions for edge and vertex weights.

`numapprox` commands for calculating polynomial approximations to functions on a given interval.

`numtheory` commands for classic number theory, primality testing, finding the $n$th prime, factoring integers, generating cyclotomic polynomials. This package also contains commands for handling convergents.

`Ore_algebra` routines for basic computations in algebras of linear operators.

`orthopoly` commands for generating various types of orthogonal polynomials; useful in differential equation solving.

`padic` commands for computing $p$-adic approximations to real numbers.

`PDEtools` tools for manipulating, solving and plotting partial differential equations.

`plots` commands for different types of specialized plots, including contour plots, two- and three-dimensional implicit plotting, plotting text, and plots in different coordinate systems.

`plottools` commands for generating and manipulating graphical objects.

`powseries` commands to create and manipulate formal power series represented in general form.

`process` the commands in this package allow you to write multi-process Maple programs under UNIX.

`simplex` commands for linear optimization using the simplex algorithm.

`stats` simple statistical manipulation of data; includes averaging, standard deviation, correlation coefficients, variance, and regression analysis.

**student**   commands for step-by-step calculus computations; including integration by parts, Simpson's rule, maximizing functions, finding extrema.

**sumtools**   commands for computing indefinite and definite sums. Includes Gosper's algorithm and Zeilberger's algorithm.

**tensor**   commands for calculating with tensors and their applications in General Relativity Theory.

**totorder**   tests for orderings between members of ordered sets.

## The Student Calculus Package

The student package helps you do step-by-step calculus computations. As an example, consider this problem: Given the function $-2/3x^2 + x$, find its derivative from first principles.

$$f'(x) = \lim_{h \to 0} \frac{f(x+h) - f(x)}{h}$$

What is the value of the derivative at $x = 0$?

```
> with(student):
```

The with command loads the student package from the Maple library. To view a list of all the commands you are reading in, replace the colon at the end of the command with a semicolon.

```
> f := x -> -2/3*x^2 + x;
```

$$f := x \to -\frac{2}{3}x^2 + x$$

```
> ( f(x+h) - f(x) )/h;
```

$$\frac{-\frac{2}{3}(x+h)^2 + h + \frac{2}{3}x^2}{h}$$

```
> Limit(%, h=0);
```

$$\lim_{h \to 0} \frac{-\frac{2}{3}(x+h)^2 + h + \frac{2}{3}x^2}{h}$$

```
> value(%);
```

$$-\frac{4}{3}x + 1$$

```
> eval(%, x=0);
```

$$1$$

To see if this seems right, plot the curve and the tangent line at $x = 0$.

```
> showtangent(f(x), x=0);
```

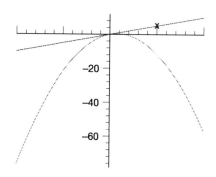

Where does this curve cross the x-axis?

```
> intercept(y=f(x), y=0);
```

$$\{y = 0, x = 0\}, \left\{y = 0, x = \frac{3}{2}\right\}$$

You can find the area under the curve between these two points, using Riemann sums.

```
> middlebox(f(x), x=0..3/2);
```

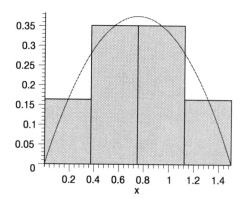

Since the result is not a good approximation, increase the number of boxes used to ten.

```
> middlebox( f(x), x=0..3/2, 10 );
```

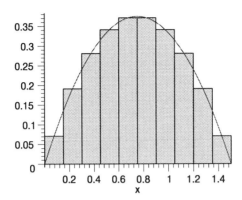

```
> middlesum( f(x), x=0..3/2, 10 );
```

$$\frac{3}{20}\left(\sum_{i=0}^{9}\left(-\frac{2}{3}\left(\frac{3}{20}i+\frac{3}{40}\right)^2+\frac{3}{20}i+\frac{3}{40}\right)\right)$$

```
> value(%);
```

$$\frac{603}{1600}$$

What is the actual value? First, use *n* boxes.

```
> middlesum( f(x), x=0..3/2, n );
```

$$\frac{3}{2}\frac{\displaystyle\sum_{i=0}^{n-1}\left(-\frac{3}{2}\frac{\left(i+\frac{1}{2}\right)^2}{n^2}+\frac{3}{2}\frac{i+\frac{1}{2}}{n}\right)}{n}$$

Then take the limit as *n* goes to ∞.

```
> Limit( %, n=infinity );
```

$$\lim_{n\to\infty}\frac{3}{2}\frac{\displaystyle\sum_{i=0}^{n-1}\left(-\frac{3}{2}\frac{\left(i+\frac{1}{2}\right)^2}{n^2}+\frac{3}{2}\frac{i+\frac{1}{2}}{n}\right)}{n}$$

```
> value(%);
```

$$\frac{3}{8}$$

Now, observe that you can obtain the same result using an integral.

```
> Int( f(x), x=0..3/2 );
```

$$\int_0^{3/2} -\frac{2}{3}x^2 + x \, dx$$

```
> value(%);
```

$$\frac{3}{8}$$

See chapter 6 for further discussions on calculus with Maple.

## The Linear Algebra Package

In linear algebra, a set of linearly independent vectors that generate the vector space is a basis. That is, you can uniquely express any element in the vector space as a linear combination of the elements of the basis.

A set of vectors $\{v_1, v_2, v_3, \ldots, v_n\}$ is linearly independent if and only if whenever

$$c_1 v_1 + c_2 v_2 + c_3 v_3 + \cdots + c_n v_n = 0$$

then

$$c_1 = c_2 = c_3 = \cdots = c_n = 0.$$

*Problem*: Determine a basis for the vector space generated by the vectors $[1, -1, 0, 1]$, $[5, -2, 3, -1]$, and $[6, -3, 3, 0]$. Express the vector $[1, 2, 3, -5]$ with respect to this basis.
*Solution*: Enter the vectors.

```
> with(linalg):
```

```
Warning, new definition for norm
Warning, new definition for trace
```

```
> v1:=vector([1,-1,0,1]):
> v2:=vector([5,-2,3,-1]):
> v3:=vector([6,-3,3,0]):
> vector_space:=stackmatrix(v1,v2,v3);
```

$$vector\_space := \begin{bmatrix} 1 & -1 & 0 & 1 \\ 5 & -2 & 3 & -1 \\ 6 & -3 & 3 & 0 \end{bmatrix}$$

If the vectors are linearly independent, then they form a basis. To test linear independence, set up the equation $c_1v_1 + c_2v_2 + c_3v_3 = 0$

$$c_1[1, -1, 0, 1] + c_2[5, -2, 3, -1] + c_3[6, -3, 3, 0] = [0, 0, 0, 0]$$

which is equivalent to

$$c_1 + 5c_2 + 6c_3 = 0$$
$$-c_1 - 2c_2 - 3c_3 = 0$$
$$3c_2 + 3c_3 = 0$$
$$c_1 - c_2 = 0$$

```
> linsolve( transpose(vector_space), [0,0,0,0] );
```

$$[-\_t_1, \; -\_t_1, \; \_t_1]$$

The vectors are linearly dependent since each is a linear product of a variable. Thus, they cannot form a basis. The command rowspace returns a basis for the vector space.

```
> b:=rowspace(vector_space);
```

$$b := \{[1, 0, 1, -1], [0, 1, 1, -2]\}$$

```
> b1:=b[1]; b2:=b[2];
```

$$b1 := [1, 0, 1, -1]$$
$$b2 := [0, 1, 1, -2]$$

```
> basis:=stackmatrix(b1,b2);
```

$$basis := \begin{bmatrix} 1 & 0 & 1 & -1 \\ 0 & 1 & 1 & -2 \end{bmatrix}$$

Express $[1, 2, 3, -5]$ in coordinates with respect to this basis.

```
> linsolve( transpose(basis), [1,2,3,-5] );
```

$$[1, 2]$$

You can find further information on this package in the ?linalg help page.

## The Matlab Package

The Matlab package allows you to call selected Matlab functions from within a Maple session.[1] Matlab is an abbreviation of **matrix laboratory**

---

[1] There is also a *Symbolic Computation Toolbox* available for Matlab that allows you to call Maple commands from within Matlab.

and is a popular numerical computation package, used extensively by engineers and other computing professionals.

In order to enable the connection between the two products, first establish the connection between the two products with

```
> Matlab[openlink]();
```

Then load the Matlab package into Maple.

```
> with(Matlab):
```

To determine the eigenvalues and eigenvectors of a matrix of integers, first define the matrix in Maple syntax.

```
> A := matrix(3, 3, [1,2,3,1,2,3,2,5,6]);
```

Then the following call to eig is made.

```
> P,W := eig(A, eigenvectors=true):
```

Notice what is to the left of the assignment operator. The (P,W) specifies that *two* outputs are to be generated and assigned to variables — the eigenvalues to W and the eigenvectors to P. This multiple assignment is available to standard Maple commands but, since existing Maple commands are designed to create a single result, is rarely used.

Let's have a look at the individual results. They are both two-dimensional hardware floating-point arrays, so we need to use the eval command to display them.

```
> eval(W);
```

```
[[9.321825,0,0],[0,-5.612673e-016,0],[0,0,-0.3218253]]
```

```
> eval(P);
```

```
[[-0.3940365,-0.9486832,-0.5567547],
    [-0.3940365,-2.758331e-017,-0.5567547],
    [-0.8303435,0.3162277,0.6164806]])
```

To translate them to standard matrices, use the convert( ,array) command.

```
> convert(P, array);
```

$$
\begin{bmatrix}
-0.39403658 & -0.94868329 & -0.55675471 \\
-0.39403658 & -2.7583318 \times 10^{-17} & -0.55675471 \\
-0.83034350 & 0.31622776 & 0.61648064
\end{bmatrix}
$$

The commands in this package can also take input in Matlab format. See the help page for Matlab for more information on acceptable input.

## The Statistics Package

The `stats` package has many commands for data analysis and manipulation, and various types of statistical plotting. It also contains a wide range
of statistical distributions.

`stats` is an example of a package which contains subpackages. Within
each subpackage are various commands grouped by functionality.

```
> with(stats);
```

$$[anova, \ describe, \ fit, \ importdata, \ random, \ statevalf,$$

$$statplots, \ transform]$$

The `stats` package works with data in *statistical lists*, which can be a
standard Maple list. A statistical list can also contain ranges and weighted
values. The difference is best shown using an example. `marks` is a standard
list,

```
> marks :=
> [64,93,75,81,45,68,72,82,76,73];
```

$$marks := [64, \ 93, \ 75, \ 81, \ 45, \ 68, \ 72, \ 82, \ 76, \ 73]$$

as is `readings`

```
> readings := [ 0.75, 0.75, .003, 1.01, .9125,
>                   .04, .83, 1.01, .874, .002 ];
```

$$readings :=$$

$$[.75, .75, .003, 1.01, .9125, .04, .83, 1.01, .874, .002]$$

which is equivalent to this statistical list.

```
> readings := [ Weight(.75, 2), .003, Weight(1.01, 2),
>                   .9125, .04, .83, .874, .002 ];
```

$$readings := [Weight(.75, 2), .003, Weight(1.01, 2),$$

$$.9125, .04, .83, .874, .002]$$

The expression `Weight(x,n)` indicates that the value $x$ appears $n$ times in
the list.

If differences less than 0.01 are so small that they are not meaningful, you can group them together, and simply give a range (using "..").

```
> readings := [ Weight(.75, 2), Weight(1.01, 2), .9125,
>               .04, .83, .874, Weight(0.002..0.003, 2) ];
```

$$readings := [\text{Weight}(.75, 2), \text{Weight}(1.01, 2), .9125, .04,$$

$$.83, .874, \text{Weight}(.002...003, 2)]$$

The describe subpackage contains commands for data analysis.

```
> describe[mean](marks);
```

$$\frac{729}{10}$$

```
> describe[range](marks);
```

$$45..93$$

```
> describe[range](readings);
```

$$.002..1.01$$

```
> describe[standarddeviation](readings);
```

$$.4038750457$$

This package contains many statistical distributions. Generate some random data using the normal distribution, group it into ranges, and then plot a histogram of the ranges.

```
> random_data:=[random[normald](50)];
```

$$random\_data := [1.175839568, -.5633641309,$$

$$.2353939952, -1.442550291, -1.079196234,$$

$$-.02201464613, -2.585278364, -.4432712806,$$

$$-1.003281481, -.02786973908, 1.526244859,$$

$$-.6051206219, .1640412457, .6530247357,$$

$$-.5410542893, 2.135025838, .1844238837,$$

$$-.6166294309, -.4486976019, .8238240523,$$

$$.2521330308, .1918301186, .8279838784,$$

$$-1.742267132, 1.123077087, -.1605619318,$$

$-1.555929034, -.7807191640, -.5186676113,$

$-.2582649678, -1.536170017, .4202060335,$

$-.5460386367, -.5234339615, .05607451436,$

$1.521577528, -1.789119833, -1.408744087,$

$-1.776317901, -.9465885589, -.1049905652,$

$1.996496081, -1.257880271, -.05641157088,$

$1.113504099, .3691334664, -.03153626578,$

$.6190037193, 1.743790472, -1.119097459]$

```
> ranges:=[-5..-2,-2..-1,-1..0,0..1,1..2,2..5];
```

$$ranges := [-5.. - 2, -2.. - 1, -1..0, 0..1, 1..2, 2..5]$$

```
> data_list:=transform[tallyinto](random_data,ranges);
```

$$data\_list := [\text{Weight}(-2.. - 1, 11), \text{Weight}(0..1, 12),$$
$$\text{Weight}(-1..0, 18), \text{Weight}(1..2, 7), 2..5, -5.. - 2]$$

```
> statplots[histogram](data_list);
```

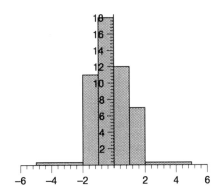

## The Linear Optimization Package

The simplex package contains commands for linear optimization, using the simplex algorithm. Linear optimization involves finding optimal solutions to equations under constraints.

An example of a classic optimization problem is the pizza delivery problem. You have four pizzas to deliver, to four different places, spread throughout the city. You want to deliver all four using as little gas as possible. You also must get to all four locations in under twenty minutes, so that the pizzas stay hot. If you can create mathematical equations representing the routes to the four places and the distances, you can find the optimal solution; that is, you can determine what route you should take in order to get to all four places in as little time and using as little gas as possible. The constraints on this particular system are that you have to deliver all four pizzas within twenty minutes of leaving the restaurant.

Here is a very small system as an example.

```
> with(simplex);
```

```
Warning, new definition for maximize
Warning, new definition for minimize
```

$$[basis, convexhull, cterm, define\_zero, display, dual,$$

$$feasible, maximize, minimize, pivot, pivoteqn, pivotvar,$$

$$ratio, setup, standardize]$$

Say you want to maximize the expression w

```
> w   := -x+y+2*z;
```

$$w := -x + y + 2z$$

subject to the constraints c1, c2, and c3.

```
> c1 := 3*x+4*y-3*z    <= 23;
```

$$c1 := 3x + 4y - 3z \le 23$$

```
> c2 := 5*x-4*y-3*z    <= 10;
```

$$c2 := 5x - 4y - 3z \le 10$$

```
> c3 := 7*x +4*y+11*z <= 30;
```

$$c3 := 7x + 4y + 11z \le 30$$

```
> maximize(w, {c1,c2,c3});
```

In this case, no answer means that Maple cannot find a solution. You can use the command feasible to determine if the set of constraints is valid.

```
> feasible({c1,c2,c3});
```

$$true$$

Try again, but this time putting an additional restriction on the solution.

```
> maximize(w, {c1,c2,c3}, NONNEGATIVE);
```

$$\left\{ x = 0, \ y = \frac{49}{8}, \ z = \frac{1}{2} \right\}$$

## 3.9  Conclusion

This chapter encompasses fundamental Maple features that will assist you greatly as you learn more complicated problem-solving methods. *Simple solve* on page 60 introduced you to solve and fsolve, and how to properly use them. Whether or not you use solve, these methods will be useful time and again.

The final sections of this chapter introduced manipulations, dsolve, and the organization of Maple and the Maple library, in attempt to give you a taste of Maple's potential. By this point in the manual, you will by no means know everything about Maple. You will, however, know enough to begin using Maple productively. You may wish to pause right now in your study of this book to work, or play, with Maple.

# Graphics

Sometimes the best way to get an understanding of a mathematical structure is to graph it in an appropriate way. Maple can produce several forms of graphs. For instance, some of its plotting capabilities include two-dimensional, two-dimensional animated, and three-dimensional graphs that you can view from any angle. Maple can handle explicit, implicit, and parametric forms, and knows several coordinate systems. Maple's flexibility allows you to conveniently manipulate graphs in many situations.

## 4.1 Graphing in Two Dimensions

When plotting an explicitly given function, $y = f(x)$, Maple needs to know the function and the domain.

```
> plot( sin(x), x=-2*Pi..2*Pi );
```

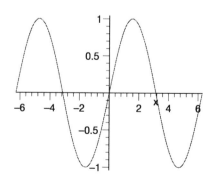

If your platform supports "smart plots", clicking on any point in the plot window reveals those particular coordinates of the plot. The menus (found on the menubar or by right-clicking on the plot itself) allow you to modify various characteristics of the plots or use many of the plotting command options listed under ?plot,options.

Maple can also graph user-defined functions.

```
> f := x -> 7*sin(x) + sin(7*x);
```

$$f := x \rightarrow 7 \sin(x) + \sin(7 x)$$

```
> plot(f(x), x=0..10);
```

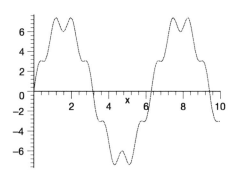

Maple allows you to focus on a given range in both the $y$- and $x$-dimension.

```
> plot(f(x), x=0..10, y=4..8);
```

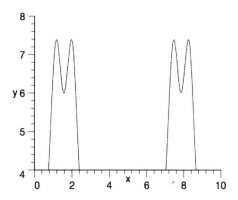

Maple can even handle infinite domains.

```
> plot( sin(x)/x, x=0..infinity);
```

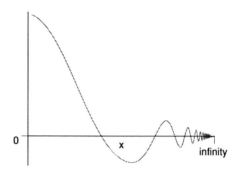

## Parametric Plots

You cannot specify some graphs explicitly; in other words, you cannot write the dependent variable as a function, $y = f(x)$. A circle is an example; for most of its $x$ values, two $y$ values exist. One solution is to make both the $x$-coordinate and the $y$-coordinate functions of some parameter, for example, $t$. The graph generated from these functions is called a *parametric plot*. Use this syntax to specify parametric plots.

```
plot( [ x-expr, y-expr, parameter=range ] )
```

That is, you plot a list containing the *x-expr*, the *y-expr*, and the name and range of the parameter. For example

```
> plot( [ t^2, t^3, t=-1..1 ] );
```

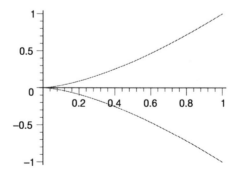

The points $(\cos t, \sin t)$ lie on a circle.

```
> plot( [ cos(t), sin(t), t=0..2*Pi ] );
```

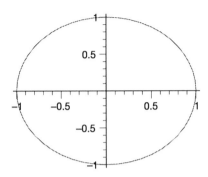

Rather than looking like a circle, the above plot resembles an ellipse because Maple, by default, scales the plot to fit the window. Here is the same plot again but with scaling=constrained. You can change the scaling using the menus or the scaling option.

```
> plot( [ cos(t), sin(t), t=0..2*Pi ], scaling=constrained );
```

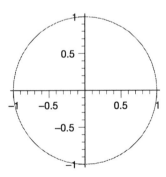

The drawback of constrained scaling is that it may obscure important details when the features in one dimension occur on a much smaller or larger scale than the others. This plot is unconstrained.

```
> plot( exp(x), x=0..3 );
```

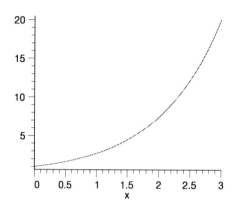

The following is the constrained version of the same plot.

```
> plot( exp(x), x=0..3, scaling=constrained);
```

## Polar Coordinates

Cartesian (ordinary) coordinates is the Maple default and is one among many ways of specifying a point in the plane; polar coordinates, $(r, \theta)$, is another option.

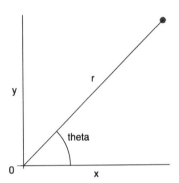

In polar coordinates, $r$ is the distance from the origin to the point, while $\theta$ is the angle between the $x$-axis and the line through the origin and the point. Maple can plot a function in polar coordinates using the polarplot command. In order to use this command, you must first employ the with command to load the plots package. The with command lists all the commands defined in the package.

```
> with(plots);
```

[*animate, animate3d, animatecurve, changecoords,*

  *complexplot, complexplot3d, conformal, contourplot,*

  *contourplot3d, coordplot, coordplot3d, cylinderplot,*

  *densityplot, display, display3d, fieldplot, fieldplot3d,*

  *gradplot, gradplot3d, implicitplot, implicitplot3d,*

  *inequal, listcontplot, listcontplot3d, listdensityplot,*

  *listplot, listplot3d, loglogplot, logplot, matrixplot,*

  *odeplot, pareto, pointplot, pointplot3d, polarplot,*

  *polygonplot, polygonplot3d, polyhedra_supported,*

  *polyhedraplot, replot, rootlocus, semilogplot,*

  *setoptions, setoptions3d, spacecurve, sparsematrixplot,*

  *sphereplot, surfdata, textplot , textplot3d, tubeplot*]

Use the following syntax to plot graphs in polar coordinates.

> polarplot( r-expr, angle=range )

In polar coordinates, you can specify the circle explicitly, namely as $r = 1$.

```
> polarplot( 1, theta=0..2*Pi, scaling=constrained );
```

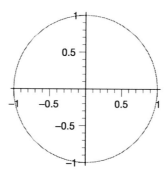

As in *Parametric Plots* on page 106, using the `scaling=constrained` option makes the circle appear round. Here is the graph of $r = \sin(3\theta)$.

```
> polarplot( sin(3*theta), theta=0..2*Pi );
```

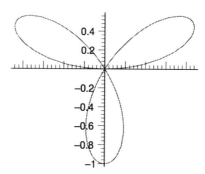

The graph of $r = \theta$ is a spiral.

```
> polarplot(theta, theta=0..4*Pi);
```

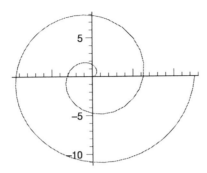

The `polarplot` command also accepts parametrized plots; that is, you can express the radius- and angle-coordinates in terms of a parameter, for example, $t$. The syntax is similar to a parametrized plot in Cartesian (ordinary) coordinates. See *Parametric Plots* on page 106.

polarplot( [ *r-expr*, *angle-expr*, *parameter=range* ] )

The equations $r = \sin(t)$ and $\theta = \cos(t)$ define the following graph.

```
> polarplot( [ sin(t), cos(t), t=0..2*Pi ] );
```

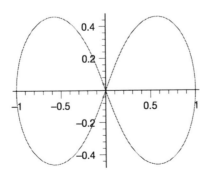

Here is the graph of $\theta = \sin(3r)$.

```
> polarplot( [ r, sin(3*r), r=0..7 ] );
```

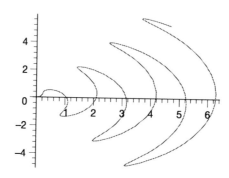

## Functions with Discontinuities

Function with discontinuities require extra attention. This function has two discontinuities, at $x = 1$ and at $x = 2$.

$$f(x) = \begin{cases} -1 & \text{if } x < 1, \\ 1 & \text{if } 1 \le x < 2, \\ 3 & \text{otherwise.} \end{cases}$$

Here is how to define $f(x)$ in Maple.

```
> f := x -> piecewise( x<1, -1, x<2, 1, 3 );
```

$$f := x \rightarrow \text{piecewise}(x < 1, -1, x < 2, 1, 3)$$

```
> plot(f(x), x=0..3);
```

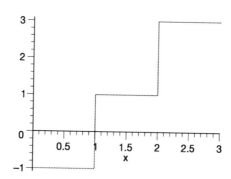

Maple draws almost vertical lines near the point of a discontinuity. The option discont=true tells Maple to watch for discontinuities.

```
> plot(f(x), x=0..3, discont=true);
```

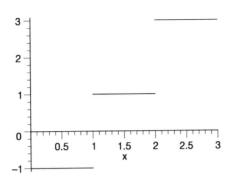

Functions with singularities, that is, those functions which become arbitrarily large at some point, constitute another special case. The function $x \mapsto 1/(x-1)^2$ has a singularity at $x = 1$.

```
> plot( 1/(x-1)^2, x=-5..6 );
```

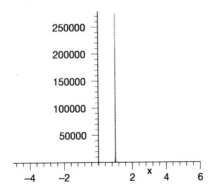

Here, all the interesting details of the graph are lost due to the tall spike at $x = 1$. The solution is to view a narrower range, perhaps from $y = -1$ to 7.

```
> plot( 1/(x-1)^2, x=-5..6, y=-1..7 );
```

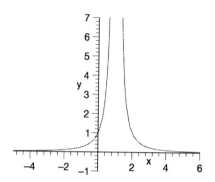

The tangent function has many singularities at $x = \frac{\pi}{2} + \pi Z$.

```
> plot( tan(x), x=-2*Pi..2*Pi );
```

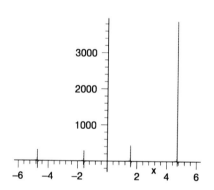

To see the details, reduce the range to $y = -4$ to 4, for example.

```
> plot( tan(x), x=-2*Pi..2*Pi, y=-4..4 );
```

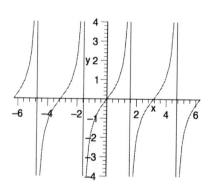

Maple draws almost vertical lines at the singularities, so you should use the `discont=true` option.

```
> plot( tan(x), x=-2*Pi..2*Pi, y=-4..4, discont=true );
```

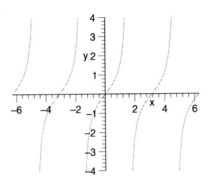

## Multiple Plots

To graph more than one function in the same plot, give `plot` a list of functions.

```
> plot( [ x, x^2, x^3, x^4 ], x=-10..10, y=-10..10 );
```

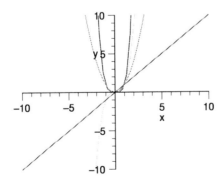

```
> f := x -> piecewise( x<0, cos(x), x>=0, 1+x^2 );
```

$$f := x \rightarrow \text{piecewise}(x < 0, \cos(x), 0 \le x, x^2 + 1)$$

```
> plot( [ f(x), diff(f(x), x), diff(f(x), x, x) ],
>    x=-2..2, discont=true );
```

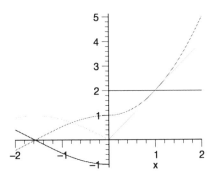

This technique also works for parametrized plots.

```
> plot( [ [ 2*cos(t), sin(t), t=0..2*Pi ],
>        [ t^2, t^3, t=-1..1 ] ] );
```

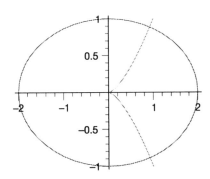

Using different line styles, such as solid, dashed, or dotted, is convenient for distinguishing between several graphs in the same plot. Maple provides the linestyle option for this. Here Maple uses linestyle=1 for the first function, $\sin(x)/x$, and linestyle=5 for the second function, $\cos(x)/x$.

```
> plot( [ sin(x)/x, cos(x)/x ], x=0..8*Pi, linestyle=[1,5] );
```

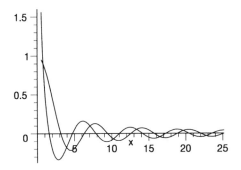

You can also change the line style using the standard menus and, if your platform supports "smart plots", the right-click pop-up menus. Similarly, specify the colors of the graphs using the color option. (You can see the effect with a color display but, in this book, the lines appear in two different shades of grey.)

```
> plot( [ [f(x), D(f)(x), x=-2..2],
>            [D(f)(x), (D@@2)(f)(x), x=-2..2] ],
>        color=[gold, plum] );
```

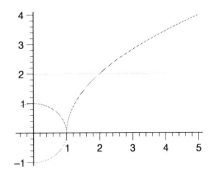

See ?plot,color for more details on colors.

## Plotting Data Points

To plot numeric data, call pointplot with the data in a list of lists of the form

$$[[x_1, y_1], [x_2, y_2], \ldots, [x_n, y_n]].$$

If the list is long, assign it to a name.

```
> data_list:=[[-2,4],[-1,1],[0, 0],[1,1],[2,4],[3,9],[4,16]];
```

$data\_list :=$

$$[[-2, 4], [-1, 1], [0, 0], [1, 1], [2, 4], [3, 9], [4, 16]]$$

```
> pointplot(data_list);
```

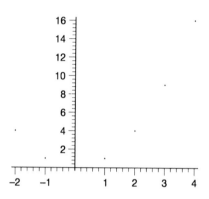

By default, Maple does not join the points with straight lines. The style=line option tells Maple to plot the lines. You can also use the menus to tell Maple to draw lines.

```
> pointplot( data_list, style=line );
```

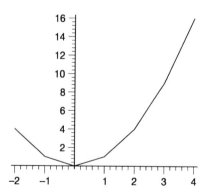

You can change the appearance of the points themselves using either the menus or the symbol option.

```
> data_list_2:=[[1,1], [2,2], [3,3], [4,4]];
```

$$data\_list\_2 := [[1, 1], [2, 2], [3, 3], [4, 4]]$$

```
> pointplot(data_list_2, style=point, symbol=cross);
```

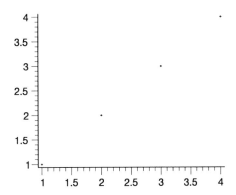

### Refining Plots

Maple uses an adaptive plotting algorithm. It calculates the value of the function or expression at a modest number of equally spaced points in the specified plotting interval. Maple then decides to compute more points within the subintervals that have a large amount of fluctuation. Occasionally, this adaptive algorithm does not cope well with a plot that has many twists and turns in small intervals.

```
> plot(sum((-1)^(i)*abs(x-i/10), i=0..30), x=-1..4);
```

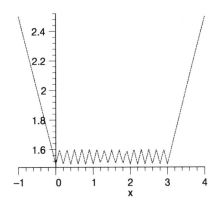

To refine this plot, you can indicate that Maple should compute more points.

```
> plot(sum((-1)^(i)*abs(x-i/10), i=0..30), x=-1..4,
>        numpoints=500);
```

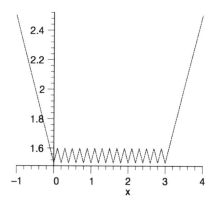

See ?plot and ?plot,options for further details and examples.

## 4.2 Graphing in Three Dimensions

You can plot a function of two variables as a surface in three-dimensional space. This allows you to visualize the function. The syntax for plot3d is similar to that for plot. Again, an explicitly given function, $z = f(x, y)$, is easiest to plot.

```
> plot3d( sin(x*y), x=-2..2, y=-2..2 );
```

You can rotate the plot by clicking in the window with your mouse and moving the bounding box. The menus allow you to change various characteristics of a plot.

As with `plot`, `plot3d` can handle user-defined functions.

```
> f := (x,y) -> sin(x) * cos(y);
```

$$f := (x, \ y) \rightarrow \sin(x)\cos(y)$$

```
> plot3d( f(x,y), x=0..2*Pi, y=0..2*Pi );
```

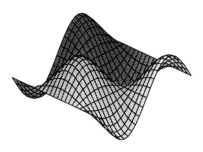

By default, Maple displays the graph as a shaded surface, but you can change this using either the menus or the `style` option. For example, `style=hidden` draws the graph as a hidden wireframe structure.

```
> plot3d( f(x,y), x=0..2*Pi, y=0..2*Pi, style=hidden );
```

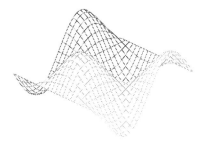

See `?plot3d,options` for a list of `style` options.

The range of the second parameter can depend on the first parameter.

```
> plot3d( sqrt(x-y), x=0..9, y=-x..x );
```

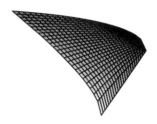

## Parametric Plots

You cannot specify some surfaces explicitly as $z = f(x, y)$; the sphere is an example of such a plot. As for two-dimensional graphs (see *Parametric Plots* on page 106), one solution is a *parametric* plot. Make the three coordinates, $x$, $y$, and $z$, functions of two parameters, for example, $s$ and $t$. You may specify parametric plots in three dimensions using the following syntax.

```
plot3d( [ x-expr, y-expr, z-expr ],
parameter1=range, parameter2=range )
```

Here are two examples.

```
> plot3d( [ sin(s), cos(s)*sin(t), sin(t) ],
>    s=-Pi..Pi, t=-Pi..Pi );
```

```
> plot3d( [ s*sin(s)*cos(t), s*cos(s)*cos(t), s*sin(t) ],
>    s=0..2*Pi, t=0..Pi );
```

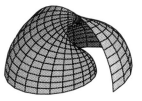

## Spherical Coordinates

The Cartesian (ordinary) coordinate system is but one of many coordinate systems in three dimensions. In the spherical coordinate system the three coordinates are the distance $r$ to the origin, the angle $\theta$ in the $xy$-plane from the $x$-axis, and the angle $\phi$ from the $z$-axis.

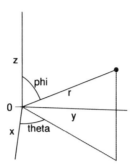

Maple can plot a function in spherical coordinates using the sphereplot command. Since the sphereplot command is defined in the plots package, you must first load the plots package. To avoid listing all the commands in the plots package, use a colon, rather than a semicolon.

```
> with(plots):
```

You can use the sphereplot command in the following manner.

> sphereplot( *r-expr*, *theta=range*, *phi=range* )

The graph of $r = (4/3)^{\theta} \sin \phi$ looks like this:

```
> sphereplot( (4/3)^theta * sin(phi),
>    theta=-1..2*Pi, phi=0..Pi );
```

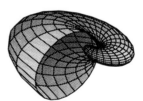

Plotting a sphere in spherical coordinates is easy: specify the radius, perhaps 1, let $\theta$ run all the way around the equator, and let $\phi$ run from the North Pole ($\phi = 0$) to the South Pole ($\phi = \pi$).

```
> sphereplot( 1, theta=0..2*Pi, phi=0..Pi,
>    scaling=constrained );
```

(See *Graphing in Two Dimensions* on page 104 for a discussion on constrained versus unconstrained plotting.)

The sphereplot command also accepts parametrized plots; that is, give the radius and both angle-coordinates in terms of two parameters, for example, $s$ and $t$. The syntax is similar to a parametrized plot in Cartesian (ordinary) coordinates. See *Parametric Plots* on page 122.

---

sphereplot( [ *r-expr*, *theta-expr*, *phi-expr* ],
*parameter1=range*, *parameter2=range* )

---

Here $r = \exp(s) + t$, $\theta = \cos(s + t)$, and $\phi = t^2$.

```
> sphereplot( [ exp(s)+t, cos(s+t), t^2 ],
>                s=0..2*Pi, t=-2..2 );
```

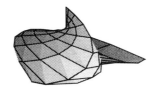

## Cylindrical Coordinates

Specify a point in the *cylindrical coordinate system* using the three coordinates $r$, $\theta$, and $z$. Here $r$ and $\theta$ are polar coordinates (see *Polar Coordinates* on page 109) in the $xy$-plane and $z$ is the usual Cartesian $z$-coordinate.

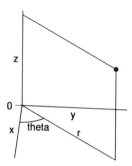

Maple plots functions in cylindrical coordinates with the `cylinderplot` command from the `plots` package.

```
> with(plots):
```

You can plot graphs in cylindrical coordinates using the following syntax.

cylinderplot( *r-expr*, *angle=range*, *z=range* )

Here is a three-dimensional version of the spiral previously shown in *Polar Coordinates* on page 109.

```
> cylinderplot( theta, theta=0..4*Pi, z=-1..1 );
```

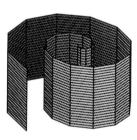

Cones are easy to plot in cylindrical coordinates: let $r$ equal $z$ and let $\theta$ run around the circle.

```
> cylinderplot( z, theta=0..2*Pi, z=0..1 );
```

cylinderplot also accepts parametrized plots. The syntax is similar to that of parametrized plots in Cartesian (ordinary) coordinates. See *Parametric Plots* on page 122.

cylinderplot( [ *r-expr*, *theta-expr*, *z-expr* ],
*parameter1=range*, *parameter2=range* )

The following is a plot of $r = st$, $\theta = s$, and $z = \cos(t^2)$.

```
> cylinderplot( [s*t, s, cos(t^2)], s=0..Pi, t=-2..2 );
```

## Refining Plots

If your plot is not as smooth or precise as you desire, tell Maple to calculate more points. The option for doing this is

grid=[*m*, *n*]

where *m* is the number of points to use for the first coordinate, and *n* is the number of points to use for the second coordinate.

```
> plot3d( sin(x)*cos(y), x=0..3*Pi, y=0..3*Pi, grid=[50,50] );
```

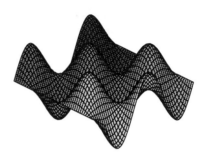

In the next example, a large number of points (100) for the first coordinate (theta) makes the spiral look smooth. However, the function does not change in the z-direction; thus, a small number of points (5) is sufficient.

```
> cylinderplot( theta, theta=0..4*Pi, z=-1..1, grid=[100,5] );
```

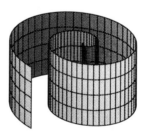

The default grid is approximately 25 by 25 points.

## Shading and Lighting Schemes

Two methods for shading a surface in a three-dimensional plot are available. In the first method, one or more distinctly colored light sources illuminate the surface. In the second method, the color of each point is a direct function of its coordinates.

Maple has a number of preselected light source configurations which give aesthetically pleasing results. You can choose from these light sources through the menus or with the `lightmodel` option. For coloring the surface directly, a number of predefined coloring functions are also available through the menus or with the `shading` option.

Simultaneous use of light sources and direct coloring may complicate the resulting coloring. Use either light sources *or* direct coloring. Here is a surface colored with `zgrayscale` shading and no lighting.

```
> plot3d( x*y^2/(x^2+y^4), x=-5..5,y=-5..5,
>      shading=zgrayscale, lightmodel=none );
```

The same surface illuminated by lighting scheme `light1` and no `shading` follows.

```
> plot3d( x*y^2/(x^2+y^4), x=-5..5,y=-5..5,
>      shading=none, lightmodel=light1 );
```

The plots appear in black and white in this book. Try them yourself to see the effects in color.

## 4.3 Animation

Graphing is an excellent way of representing information. However, static plots do not always emphasize certain graphical behavior, such as the deformation of a bouncing ball, as well as their animated counterparts.

A Maple animation is a sequence of plot frames displayed rapidly one after the other, similar to the action of movie frames. The two commands used for animations, animate and animate3d, are defined in the plots package. Remember to load this package using the with command before you first use any of its commands.

### Animation in Two Dimensions

You can specify a two-dimensional animation using this syntax.

animate( *y-expr*, *x=range*, *time=range* )

The following is an example of an animation.

```
> with(plots):
> animate( sin(x*t), x=-10..10, t=1..2 );
```

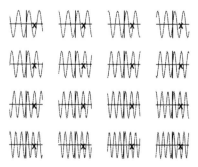

To play an animation you must first select it by clicking on it. Then choose Play from the Animation menu.

By default, a two dimensional animation consists of sixteen plots (frames). If the motion is ragged, tell Maple to make more frames.

```
> animate( sin(x*t), x=-10..10, t=1..2, frames=50);
```

The usual `plot` options are also available.

```
> animate( sin(x*t), x=-10..10, t=1..2,
>    frames=50, numpoints=100 );
```

You can plot any two-dimensional animation as a three-dimensional static plot. For example, try plotting the animation of $\sin(xt)$ above as a surface.

```
> plot3d( sin(x*t), x=-10..10, t=1..2, grid=[50,100],
>    orientation=[135,45], axes=boxed, style=hidden );
```

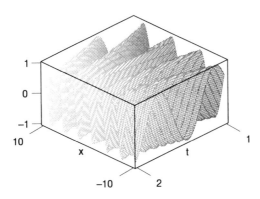

Whether you prefer an animation or a plot is a matter of taste and also depends on the concepts that the animation or plot is supposed to convey.

Animating parametrized graphs is also possible. (See *Parametric Plots* on page 106.)

```
> animate( [ a*cos(u), sin(u), u=0..2*Pi ], a=0..2 );
```

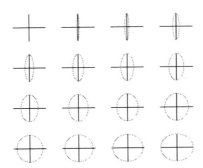

The coords option tells animate to use a coordinate system other than the Cartesian (ordinary) system.

```
> animate( theta*t, theta=0..8*Pi, t=1..4, coords=polar );
```

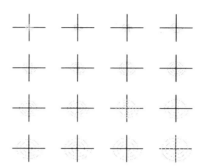

Displaying animations in a book is difficult because still pictures cannot convey the same graphical behavior as those in a movie. Therefore, you should enter these suggested commands into Maple yourself to get a better understanding.

## Animation in Three Dimensions

Use animate3d to animate surfaces in three dimensions. You can use the animate3d command as follows.

animate3d( *z-expr*, *x=range*, *y=range*, *time=range* )

The following is an example of a three-dimensional animation.

```
> animate3d( cos(t*x)*sin(t*y),
>              x=-Pi..Pi, y=-Pi..Pi, t=1..2 );
```

By default, a three-dimensional animation consists of eight plots. As for two-dimensional animations, the frames option tells Maple how many frames to make.

```
> animate3d( cos(t*x)*sin(t*y), x=-Pi..Pi, y=-Pi..Pi, t=1..2,
>     frames=16 );
```

*Parametric Plots* on page 122 describes three-dimensional parametrized plots. You can also animate these.

```
> animate3d( [s*time, t-time, s*cos(t*time)],
>     s=1..3, t=1..4, time=2..4, axes=boxed);
```

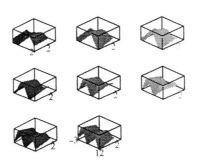

To animate a function in a coordinate system other than the Cartesian, use the coords option. For spherical coordinates use, coords=spherical.

```
> animate3d( (1.3)^theta * sin(t*phi), theta=-1..2*Pi,
>     phi=0..Pi, t=1..8, coords=spherical );
```

For cylindrical coordinates, use coords=cylindrical.

```
> animate3d( sin(theta)*cos(z*t), theta=1..3, z=1..4,
>     t=1/4..7/2, coords=cylindrical );
```

See ?plots,changecoords for a list of the coordinate systems Maple knows.

Please note that computing many frames sometimes requires a lot of time and memory.

## 4.4  Annotating Plots

Adding text annotation to plots is possible in a variety of ways. The option title prints the specified title in the plot window, centered and near the top.

```
> plot( sin(x), x=-2*Pi..2*Pi, title="Plot of Sine" );
```

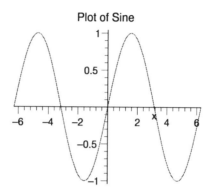

Note that when specifying the title you must place a double quotes (")
at both ends of the text. This is very important. Maple uses double quotes
to delimit strings. It considers whatever appears between double quotes to
be a piece of text that it should not process further. You can specify the font,
style, and size of the title with the titlefont option. See ?plot,options
or ?plot3d,options.

```
> with(plots):
> sphereplot( 1, theta=0..2*Pi, phi=0..Pi,
>    scaling=constrained, title="The Sphere",
>    titlefont=[HELVETICA, BOLD, 24] );
```

# The Sphere

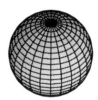

The labels option allows you to specify the labels on the axes, and the
labelsfont option gives you control over the font and style of the labels.
Note that the labels do not have to match the variables in the expression
you are plotting.

# Color Plates

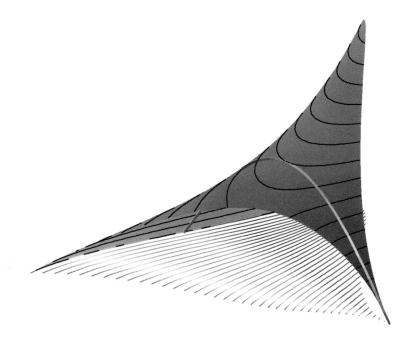

Plate 1: Solution to a Partial Differential Equation

Plate 2: An Apple in a Cage

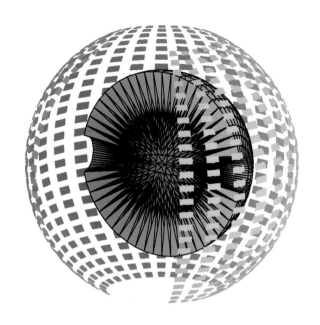

Plate 3: A Checkered Cut-Away

Plate 4: A Colorful Tree

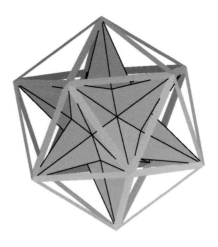

Plate 5: Nested Polyhedra

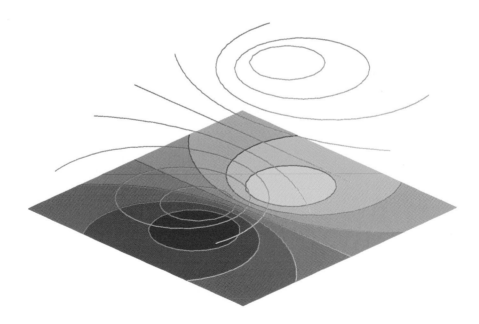

Plate 6: Projections of Contours

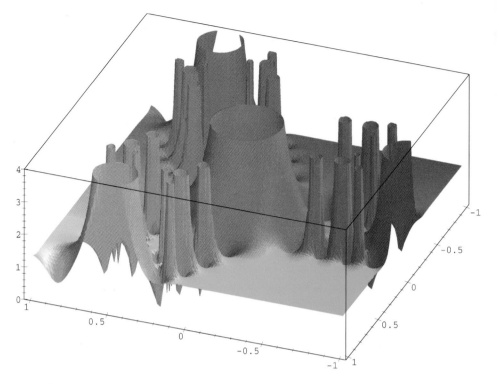

Plate 7: A Complex Surface

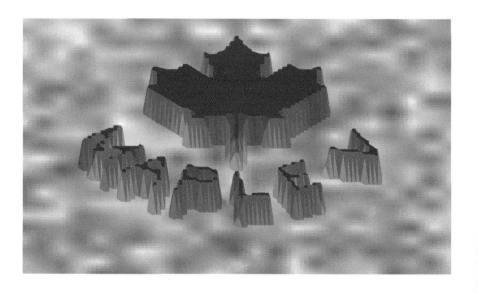

Plate 8: A Creative Endeavor

```
> plot( x^2, x=0..4, labels=["time", "velocity"] );
```

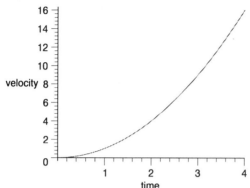

You can print labels only if your plot displays axes. For three-dimensional graphs, the default is no axes, so ask for axes explicitly.

```
> plot3d( sin(x*y), x=-1..1, y=-1..1,
>       labels=["length", "width", "height"], axes=FRAMED );
```

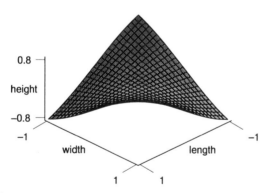

See also *Placing Text in Plots* on page 137.

## 4.5 Composite Plots

Maple allows you to display several plots simultaneously, after assigning names to the individual plots. Since plot structures are usually rather large, end the assignments with colons (rather than semicolons).

```
> my_plot := plot( sin(x), x=-10..10 ):
```

Now you can save the plot for future use, as you would any other expression. Exhibit the plot using the display command defined in the plots package.

```
> with(plots):
> display( my_plot );
```

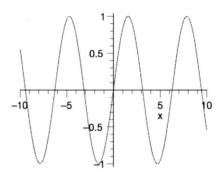

The display command can draw several plots at the same time. Simply give a list of plots.

```
> a := plot( [ sin(t), exp(t)/20, t=-Pi..Pi ] ):
> b := polarplot( [ sin(t), exp(t), t=-Pi..Pi ] ):
> display( [a,b] );
```

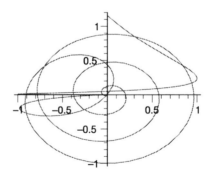

This technique allows you to display plots of different types in the same axes. You can also display three-dimensional plots, even animations.

```
> c := sphereplot( 1, theta=0..2*Pi, phi=0..Pi ):
> d := cylinderplot( 0.5, theta=0..2*Pi, z=-2..2 ):
```

```
> display( [c,d], scaling=constrained );
```

```
> e := animate( m*x, x=-1..1, m=-1..1 ):
> display( [b,e] );
```

If you display two or more animations together, make sure that they have the same number of frames.

```
> f := animate3d( sin(x+y+t), x=0..2*Pi, y=0..2*Pi, t=0..5,
>       frames=20 ):
> g := animate3d( t, x=0..2*Pi, y=0..2*Pi, t=-1.5..1.5,
>       frames=20):
> display( [f,g] );
```

## Placing Text in Plots

The `title` and `labels` options to the plotting commands allow you to put titles and labels on your graphs. The `textplot` and `textplot3d` commands give more flexibility by allowing you to specify the exact positions of the text. The `plots` package defines these two commands.

```
> with(plots):
```

You may use `textplot` and `textplot3d` as follows.

```
textplot( [ x-coord, y-coord, "text" ] );
textplot3d( [ x-coord, y-coord, z-coord, "text"] );
```

For example

```
> a := plot( sin(x), x=-Pi..Pi ):
> b := textplot( [ Pi/2, 1, "Local Maximum" ] ):
> c := textplot( [ -Pi/2, -1, "Local Minimum" ] ):
```

```
> display( [a,b,c], title="The Sine Curve" );
```

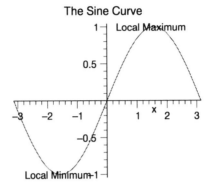

See ?plots,textplot for details on how to fine tune the placing of the text. Use the font option to specify which font textplot and textplot3d should use.

```
> d := plot3d( x^2-y^2, x=-1..1, y=-1..1 ):
> e := textplot3d( [0, 0, 0, "A Saddle Point"],
>       font=[HELVETICA, OBLIQUE, 22], color=black ):
> display( [d,e], orientation=[68,45] );
```

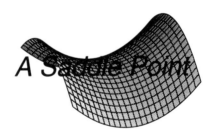

## 4.6 Special Types of Plots

The plots package contains many routines for producing special types of graphics.

Here is a variety of examples. For further explanation of a particular plot command, see ?plots,*command*.

```
> with(plots):
```

Plot implicitly defined functions using implicitplot.

```
> implicitplot( x^2+y^2=1, x=-1..1, y=-1..1,
>       scaling=constrained );
```

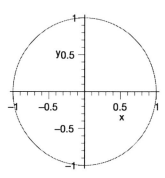

Below is a plot of the region satisfying the inequalities $x + y < 5, 0 < x$, and $x \leq 4$.

```
> inequal( {x+y<5, 0<x, x<=4}, x=-1..5, y=-10..10,
>       optionsexcluded=(color=yellow) );
```

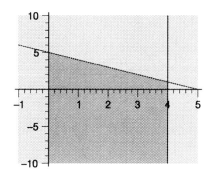

Here the vertical axis has a logarithmic scale.

```
> logplot( 10^x, x=0..10 );
```

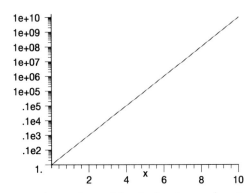

A `semilogplot` has a logarithmic horizontal axis.

```
> semilogplot( 2^(sin(x)), x=1..10 );
```

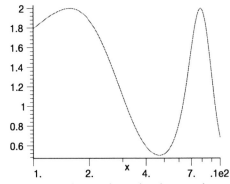

Maple can also create plots where both axes have logarithmic scales.

```
> loglogplot( x^17, x=1..7 );
```

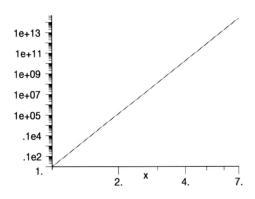

In a `densityplot`, lighter shading indicates a larger function value.

```
> densityplot( sin(x*y), x=-1..1, y=-1..1 );
```

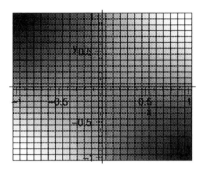

Along the following curves, $\sin(xy)$ is constant, as in a topographical map.

```
> contourplot(sin(x*y),x=-10..10,y=-10..10);
```

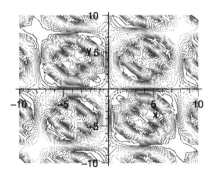

A rectangular grid in the complex plane becomes the following graph when you map it by $z \mapsto z^2$.

```
> conformal( z^2, z=0..2+2*I );
```

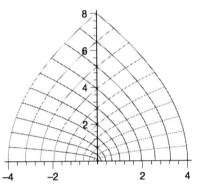

The `fieldplot` command draws the given vector for many values of x and y; that is, it plots a vector field, such as a magnetic field.

```
> fieldplot( [y*cos(x*y), x*cos(x*y)], x=-1..1, y=-1..1);
```

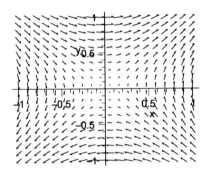

Maple can draw curves in three-dimensional space.

```
> spacecurve( [cos(t),sin(t),t], t=0..4*Pi );
```

Here Maple inflates the above spacecurve to form a tube.

```
> tubeplot( [cos(t),sin(t),t], t=0..4*Pi, radius=0.5 );
```

The `matrixplot` command plots the values of a matrix.

```
> A := linalg[hilbert](8):
> B := linalg[toeplitz]([1,2,3,4,-4,-3,-2,-1]):
> matrixplot( evalm(A+B), heights=histogram, axes=frame,
>     gap=0.25, style=patch);
```

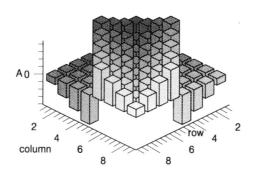

Below is a demonstration of a root locus plot.

```
> rootlocus( (s^5-1)/(s^2+1), s, -5..5, style=point,
>      adaptive=false );
```

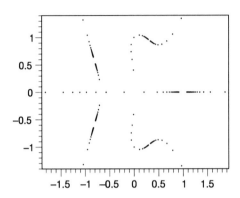

Typing ?plots provides you with a listing of other available plot types.

## 4.7  Manipulating Graphical Objects

The plottools package contains commands for creating graphical objects and manipulating their plots. Load this package before using the commands.

```
> with(plottools):
```

The objects in the plottools package do not display by themselves; you must use the display command, defined in the plots package, which you must load as well.

```
> with(plots):
```

Now you are ready for an example.

```
> display( dodecahedron(), scaling=constrained, style=patch );
```

Give an object a name.

```
> s1 := sphere( [3/2,1/4,1/2], 1/4, color=red):
```

Note that the assignment ends with a colon; if you use a semicolon, Maple prints out a large plot structure. Again, you must use display to see the plot.

```
> display( s1, scaling=constrained );
```

Place a second sphere in the picture and ask for axes.

```
> s2 := sphere( [3/2,-1/4,1/2], 1/4, color=red):
```

```
> display( [s1, s2], axes=normal, scaling=constrained );
```

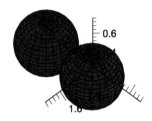

You can also make cones with the plottools package.

```
> c := cone([0,0,0], 1/2, 2, color=khaki):
> display( c, axes=normal );
```

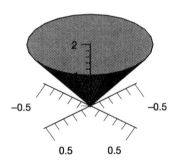

Experiment using Maple's object rotation capabilities.

```
> c2 := rotate( c, 0, Pi/2, 0 ):
> display( c2, axes=normal );
```

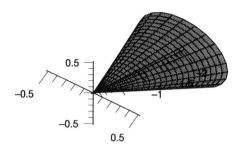

Translating objects is yet another option.

```
> c3 := translate( c2, 3, 0, 1/4 ):
> display( c3, axes=normal );
```

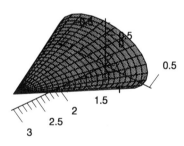

The hemisphere command makes a hemisphere. You may specify the radius and the coordinates of the center. Otherwise, leave an empty set of parentheses to accept the defaults.

```
> cup := hemisphere():
```

```
> display( cup );
```

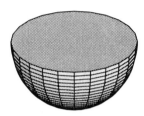

```
> cap := rotate( cup, Pi, 0, 0 ):
> display( cap );
```

The sides of the dodecahedron mentioned earlier in this section are all pentagons. If you raise the midpoint of each pentagon using the `stellate` command, the term for the resulting object is *stellated* dodecahedron.

```
> a := stellate( dodecahedron() ):
```

```
> display( a, scaling=constrained, style=patch );
```

```
> stelhs := stellate(cap, 2):
> display( stelhs );
```

Instead of stellating the dodecahedron, you may cut out, for example, the inner three quarters of each pentagon.

```
> a := cutout( dodecahedron(), 3/4 ):
```

```
> display(  a, scaling=constrained, orientation=[45, 30] );
```

```
> hedgehog := [s1, s2, c3, stelhs]:
> display( hedgehog, scaling=constrained,
>     style=patchnogrid );
```

## 4.8  Conclusion

This chapter examined Maple's two- and three-dimensional plotting capabilities, involving explicitly, parametricly, and implicitly given functions. Cartesian, polar, spherical, and cylindrical are a few of the many coordinate systems that Maple can handle. Furthermore, you can animate a graph and shade it in a variety of ways for a clearer understanding of its nature.

Use the commands found in the plots package to display various graphs of functions and expressions. The commands within the plottools package create and manipulate objects. Such commands, for instance, allow you to translate, rotate, and even stellate a graphical entity.

# Evaluation and Simplification

In Maple, a significant amount of time and effort is spent manipulating expressions. Expression manipulation is done for many reasons, from converting output expressions into a familiar form to check answers, to converting expressions into a specific form needed by certain Maple routines.

The issue of simplification is surprisingly difficult in symbolic mathematics. What is "simple" in one context may not be in another context—each individual context can have its own definition of a "simple" form.

Maple provides a set of tools for working with expressions, both for mathematical manipulations and structural manipulations. Mathematical manipulations are those that correspond to some kind of standard mathematical process, for example factoring a polynomial, or rationalizing the denominator of a rational expression. Structural manipulations are tools Maple provides to let you access and modify parts of the Maple data structures which represent expressions and other types of objects.

## 5.1 Mathematical Manipulations

Solving equations by hand usually involves performing a sequence of algebraic manipulations. You can also perform these steps using Maple.

```
> eq := 4*x + 17 = 23;
```

$$eq := 4\,x + 17 = 23$$

Here, you must subtract 17 from both sides of the equation. To do so, subtract the equation 17=17 from eq. Make sure to put parentheses around the unnamed equation.

```
> eq - ( 17 = 17 );
```

$$4x = 6$$

Now divide through by 4. Note that you don't have to use 4=4 in this case.

```
> % / 4;
```

$$x = \frac{3}{2}$$

The following sections focus on many more sophisticated manipulations.

## Expanding Polynomials as Sums

Sums are generally easier to comprehend than products, so you may find it useful to expand a polynomial as a sum of products. The expand command has this capability.

```
> poly := (x+1)*(x+2);
```

$$poly := (x + 1)(x + 2)$$

```
> expand( poly );
```

$$x^2 + 3x + 2$$

The expand command expands the numerator of a rational expression.

```
> expand( (x+1)*(y^2-2*y+1) / z / (y-1) );
```

$$\frac{x\,y^2}{z\,(y-1)} - 2\frac{x\,y}{z\,(y-1)} + \frac{x}{z\,(y-1)} + \frac{y^2}{z\,(y-1)}$$

$$- 2\frac{y}{z\,(y-1)} + \frac{1}{z\,(y-1)}$$

Use the normal command to cancel common factors; see *Factored Normal Form* on page 159.

The expand command also knows expansion rules for many standard mathematical functions.

```
> expand( sin(2*x) );
```

$$2\sin(x)\cos(x)$$

```
> ln( abs(x^2)/(1+abs(x)) );
```

$$\ln\left(\frac{|x|^2}{1 + |x|}\right)$$

```
> expand(%);
```

$$2 \ln (|x|) - \ln (1 + |x|)$$

The combine command knows the same rules but applies them in the opposite direction. See *Combining Terms* on page 158.

You can specify subexpressions that you do *not* want to expand, as an argument to expand.

```
> expand( (x+1)*(y+z) );
```

$$x y + x z + y + z$$

```
> expand( (x+1)*(y+z), x+1 );
```

$$(x + 1) y + (x + 1) z$$

You may expand an expression over a special domain.

```
> poly := (x+2)^2*(x-2);
```

$$poly := (x + 2)^2 (x - 2)$$

```
> expand( poly );
```

$$x^3 + 2 x^2 - 4 x - 8$$

```
> % mod 3;
```

$$x^3 + 2 x^2 + 2 x + 1$$

However, using the Expand command is more efficient.

```
> Expand( poly ) mod 3;
```

$$x^3 + 2 x^2 + 2 x + 1$$

When you use Expand with mod, Maple performs all intermediate calculations in modulo arithmetic. You can also write your own expand subroutines; see ?expand for more details.

## Collecting the Coefficients of Like Powers

An expression like $x^2 + 2x + 1 - ax + b - cx^2$ may be easier to read if you collect together the coefficients of $x^2$, $x$, and the constant terms, using the collect command.

```
> collect( x^2 + 2*x + 1 - a*x + b - c*x^2, x );
```

$$(1 - c) x^2 + (2 - a) x + b + 1$$

The second argument to the collect command specifies on which variable it should collect.

```
> poly := x^2 + 2*y*x - 3*y + y^2*x^2;
```

$$poly := x^2 + 2\,y\,x - 3\,y + y^2\,x^2$$

```
> collect( poly, x );
```

$$(1 + y^2)\,x^2 + 2\,y\,x - 3\,y$$

```
> collect( poly, y );
```

$$y^2\,x^2 + (2\,x - 3)\,y + x^2$$

You can collect on either variables or unevaluated function calls.

```
> trig_expr := sin(x)*cos(x) + sin(x) + y*sin(x);
```

$$trig\_expr := \sin(x)\cos(x) + \sin(x) + y\sin(x)$$

```
> collect( trig_expr, sin(x) );
```

$$(\cos(x) + 1 + y)\sin(x)$$

```
> DE := diff(f(x),x,x)*sin(x) - diff(f(x),x)*sin(f(x))
>      + sin(x)*diff(f(x),x) + sin(f(x))*diff(f(x),x,x);
```

$$DE := \left(\frac{\partial^2}{\partial x^2}\,f(x)\right)\sin(x) - \left(\frac{\partial}{\partial x}\,f(x)\right)\sin(f(x))$$

$$+ \sin(x)\left(\frac{\partial}{\partial x}\,f(x)\right) + \sin(f(x))\left(\frac{\partial^2}{\partial x^2}\,f(x)\right)$$

```
> collect( DE, diff );
```

$$(-\sin(f(x)) + \sin(x))\left(\frac{\partial}{\partial x}\,f(x)\right)$$

$$+ (\sin(x) + \sin(f(x)))\left(\frac{\partial^2}{\partial x^2}\,f(x)\right)$$

You cannot collect on sums or products.

```
> big_expr := z*x*y + 2*x*y + z;
```

$$big\_expr := z\,x\,y + 2\,y\,x + z$$

```
> collect( big_expr, x*y );
```

```
Error, (in collect) cannot collect, y*x
```

Instead, make a substitution before you collect. In the above case, substituting a dummy name for x*y, and then collecting on the dummy name achieves what you want.

```
> subs( x=xyprod/y, big_expr );
```

$$z \, xyprod + 2 \, xyprod + z$$

```
> collect( %, xyprod );
```

$$(z + 2) \, xyprod + z$$

```
> subs( xyprod=x*y, % );
```

$$(z + 2) \, y \, x + z$$

*Substitution* on page 183 explains the use of the subs command.

If you are collecting coefficients of more than one variable at a time, two options are available: recursive form, or distributed form. Recursive form initially collects in the first specified variable, and then in the next, and so on. The default is the recursive form.

```
> poly := x*y + z*x*y + y*x^2 - z*y*x^2 + x + z*x;
```

$$poly := y \, x + z \, x \, y + y \, x^2 - z \, y \, x^2 + x + z \, x$$

```
> collect( poly, [x,y] );
```

$$(1 - z) \, y \, x^2 + ((1 + z) \, y + 1 + z) \, x$$

Distributed form collects the coefficients of all variables at the same time.

```
> collect( poly, [x,y], distributed );
```

$$(1 + z) \, x + (1 + z) \, y \, x + (1 - z) \, y \, x^2$$

The collect command does not sort the terms; use the sort command for that. See *Sorting Algebraic Expressions* on page 163.

## Factoring Polynomials and Rational Functions

You may want to write a polynomial as a product of terms of smallest possible degree. Use the factor command to factor polynomials.

```
> factor( x^2-1 );
```

$$(x - 1) \, (x + 1)$$

```
> factor( x^3+y^3 );
```

$$(x + y) \, (x^2 - y \, x + y^2)$$

You can also factor rational functions. The factor command factors both the numerator and the denominator, and then removes common terms.

```
> rat_expr := (x^16 - y^16) / (x^8 - y^8);
```

$$rat\_expr := \frac{x^{16} - y^{16}}{x^8 - y^8}$$

```
> factor( rat_expr );
```

$$x^8 + y^8$$

```
> rat_expr := (x^16 - y^16) / (x^7 - y^7);
```

$$rat\_expr := \frac{x^{16} - y^{16}}{x^7 - y^7}$$

```
> factor(rat_expr);
```

$$\frac{(x + y)(x^2 + y^2)(x^4 + y^4)(x^8 + y^8)}{x^6 + y\,x^5 + y^2\,x^4 + y^3\,x^3 + y^4\,x^2 + y^5\,x + y^6}$$

**Specifying the Algebraic Number Field**  The factor command factors a polynomial over the ring implied by the coefficients. The following polynomial has integer coefficients, so the terms in the factored form have integer coefficients

```
> poly := x^5 - x^4 - x^3 + x^2 - 2*x + 2;
```

$$poly := x^5 - x^4 - x^3 + x^2 - 2\,x + 2$$

```
> factor( poly );
```

$$(x - 1)(x^2 - 2)(x^2 + 1)$$

In this next example, the coefficients include $\sqrt{2}$. Note the differences in the result.

```
> expand( sqrt(2)*poly );
```

$$\sqrt{2}\,x^5 - \sqrt{2}\,x^4 - \sqrt{2}\,x^3 + \sqrt{2}\,x^2 - 2\,\sqrt{2}\,x + 2\,\sqrt{2}$$

```
> factor( % );
```

$$\sqrt{2}\,(x^2 + 1)(x - \sqrt{2})(x + \sqrt{2})(x - 1)$$

You can explicitly extend the coefficient field by giving a second argument to factor.

```
> poly := x^4 - 5*x^2 + 6;
```

$$poly := x^4 - 5\,x^2 + 6$$

```
> factor( poly );
```

$$(x^2 - 2)(x^2 - 3)$$

```
> factor( poly, sqrt(2) );
```

$$(x^2 - 3)(x - \sqrt{2})(x + \sqrt{2})$$

```
> factor( poly, { sqrt(2), sqrt(3) } );
```

$$(x - \sqrt{2})(x + \sqrt{2})(x - \sqrt{3})(x + \sqrt{3})$$

You can also specify the extension using RootOf. Here RootOf(x^2-2) represents any solution to $x^2 - 2 = 0$, that is either $\sqrt{2}$ or $-\sqrt{2}$.

```
> factor( poly, RootOf(x^2-2) );
```

$$(x^2 - 3)(x - \text{RootOf}(\_Z^2 - 2))(x + \text{RootOf}(\_Z^2 - 2))$$

See ?evala for more on how to calculate in an algebraic number field.

**Factoring in Special Domains**  Use the Factor command to factor an expression over the integers modulo $p$ for some prime $p$. The syntax is similar to that of the Expand command.

```
> Factor( x^2+3*x+3 ) mod 7;
```

$$(x + 4)(x + 6)$$

The Factor command also allows algebraic field extensions.

```
> Factor( x^3+1 ) mod 5;
```

$$(x^2 + 4x + 1)(x + 1)$$

```
> Factor( x^3+1, RootOf(x^2+x+1) ) mod 5;
```

$$(x + 4\,\text{RootOf}(\_Z^2 + \_Z + 1) + 4)$$

$$(x + \text{RootOf}(\_Z^2 + \_Z + 1))(x + 1)$$

For details about the algorithm used, factoring multivariate polynomials, or factoring polynomials over an algebraic number field, see ?Factor.

## Removing Rational Exponents

Rational expressions are generally considered in nicer form if they have no fractional exponents in the denominator. The rationalize command removes roots from the denominator of a rational expression by multiplying through by a suitable factor.

```
> 1 / ( 2 + root[3](2) );
```

$$\frac{1}{2 + 2^{1/3}}$$

> rationalize( % );

$$\frac{2}{5} - \frac{1}{5} 2^{1/3} + \frac{1}{10} 2^{2/3}$$

> (x^2+5) / (x + x^(5/7));

$$\frac{x^2 + 5}{x + x^{5/7}}$$

> rationalize( % );

$$\frac{(x^2 + 5)(x^{6/7} - x^{12/7} - x^{4/7} + x^{10/7} + x^{2/7} - x^{8/7} + x^2)}{x^3 + x}$$

The result of `rationalize` is often larger than the original.

## Combining Terms

The `combine` command applies a number of transformation rules for various mathematical functions.

> combine( sin(x)^2 + cos(x)^2 );

$$1$$

> combine( sin(x)*cos(x) );

$$\frac{1}{2} \sin(2x)$$

> combine( exp(x)^2 * exp(y) );

$$e^{(2x+y)}$$

> combine( (x^a)^2 );

$$x^{(2a)}$$

To see how `combine` arrives at the result, give `infolevel[combine]` a positive value.

> infolevel[combine] := 1;

$$infolevel_{combine} := 1$$

> expr := Int(1, x) + Int(x^2, x);

$$expr := \int 1\, dx + \int x^2\, dx$$

> combine( expr );
combine:    combining with respect to    combine/Int
combine:    combining with respect to    combine/linear

```
combine:    combining with respect to    combine/int
combine:    combining with respect to    combine/linear
combine:    combining with respect to    combine/range
combine:    combining with respect to    combine/Int
combine:    combining with respect to    combine/linear
combine:    combining with respect to    combine/range
combine:    combining with respect to    combine/int
combine:    combining with respect to    combine/linear
combine:    combining with respect to    combine/range
combine:    combining with respect to    combine/Int
combine:    combining with respect to    combine/linear
combine:    combining with respect to    combine/range
combine:    combining with respect to    combine/range
combine:    combining with respect to    combine/cmbplus
combine:    combining with respect to    combine/cmbplus
combine:    combining with respect to    combine/cmbplus
```

$$\int x^2 + 1 \, dx$$

The expand command applies most of these transformation rules in the other direction. See *Expanding Polynomials as Sums* on page 152.

### Factored Normal Form

If an expression contains fractions, you may find it useful to turn the expression into one large fraction, and cancel common factors in the numerator and denominator. The normal command performs this process, which often leads to simpler expressions.

```
> normal( x + 1/x );
```

$$\frac{x^2 + 1}{x}$$

```
> expr := x/(x+1) + 1/x + 1/(1+x);
```

$$expr := \frac{x}{x+1} + \frac{1}{x} + \frac{1}{x+1}$$

```
> normal( expr );
```

$$\frac{x+1}{x}$$

```
> expr := (x^2 - y^2) / (x-y)^3;
```

$$expr := \frac{x^2 - y^2}{(x - y)^3}$$

```
> normal( expr );
```

$$\frac{x+y}{(x-y)^2}$$

The normal command puts the numerator in expanded form.

```
> expr := (x - 1/x) / (x-2);
```

$$expr := \frac{x - \dfrac{1}{x}}{x - 2}$$

```
> normal( expr );
```

$$\frac{x^2 - 1}{x\,(x - 2)}$$

Use the second argument expanded if you want normal to also expand the denominator.

```
> normal( expr, expanded );
```

$$\frac{x^2 - 1}{x^2 - 2\,x}$$

The normal command acts recursively over functions, sets, and lists.

```
> normal( [ expr, exp(x+1/x) ] );
```

$$\left[ \frac{x^2 - 1}{x\,(x - 2)},\ e^{\left(\frac{x^2+1}{x}\right)} \right]$$

```
> big_expr := sin( (x*(x+1)-x)/(x+2) )^2
>            + cos( (x^2)/(-x-2) )^2;
```

$$big\_expr := \sin\left(\frac{(x+1)\,x - x}{x+2}\right)^2 + \cos\left(\frac{x^2}{-x-2}\right)^2$$

```
> normal( big_expr );
```

$$\sin\left(\frac{x^2}{x+2}\right)^2 + \cos\left(\frac{x^2}{x+2}\right)^2$$

Note from the last example above that normal does not know how to simplify trigonometric expressions, only rational functions. The combine command can simplify many mathematical functions. See *Combining Terms* on page 158.

**A Special Case**  Because it expands the numerator in the result, `normal` is less helpful in cases where the expanded form of the numerator is not as simple as the factored form.

```
> expr := (x^25-1) / (x-1);
```

$$expr := \frac{x^{25} - 1}{x - 1}$$

```
> normal( expr );
```

$$x + x^2 + x^{16} + x^4 + x^3 + x^5 + x^7 + x^{22} + x^{23} + x^{21} + x^{20} + x^{19}$$
$$+ x^{18} + x^{17} + x^{15} + x^{14} + x^{13} + x^{12} + x^{11} + x^{10} + x^9 + x^8$$
$$+ x^6 + x^{24} + 1$$

To cancel out the common $(x - 1)$ term from the numerator and the denominator without expanding the numerator, use `factor`. See *Factoring Polynomials and Rational Functions* on page 155.

```
> factor(expr);
```

$$(x^4 + x^3 + x^2 + x + 1)(x^{20} + x^{15} + x^{10} + x^5 + 1)$$

Although the `normal` command is very useful, normalizing expressions can be computationally intensive and may not always produce a simpler form.

## Simplifying Expressions

The results of Maple's simplification calculations can be very complicated. The `simplify` command applies a list of manipulations which tries to find a simpler expression.

```
> expr := 4^(1/2) + 3;
```

$$expr := \sqrt{4} + 3$$

```
> simplify( expr );
```

$$5$$

```
> expr := cos(x)^5 + sin(x)^4 + 2*cos(x)^2
>    - 2*sin(x)^2 - cos(2*x);
```

$$expr :=$$
$$\cos(x)^5 + \sin(x)^4 + 2\cos(x)^2 - 2\sin(x)^2 - \cos(2x)$$

```
> simplify( expr );
```

$$\cos(x)^5 + \cos(x)^4$$

Simplification rules are known for trigonometric expression, logarithmic and exponential expressions, radical expressions, expressions with powers, RootOf expressions, and various special functions.

If you specify a particular simplification rule as an argument to the simplify command, then it uses only that simplification rule (or that class of rules).

```
> expr := ln(3*x) + sin(x)^2 + cos(x)^2;
```

$$expr := \ln(3\,x) + \sin(x)^2 + \cos(x)^2$$

```
> simplify( expr, trig );
```

$$\ln(3\,x) + 1$$

```
> simplify( expr, ln );
```

$$\ln(3) + \ln(x) + \sin(x)^2 + \cos(x)^2$$

```
> simplify( expr );
```

$$\ln(3) + \ln(x) + 1$$

See ?simplify for a list of built-in simplification rules.

## Simplification with Assumptions

Maple may refuse to perform an obvious simplification because, although you know that a variable has special properties, Maple treats the variable in a more general way.

```
> expr := sqrt( (x*y)^2 );
```

$$expr := \sqrt{x^2\,y^2}$$

```
> simplify( expr );
```

$$\sqrt{x^2\,y^2}$$

The option assume=*property* tells simplify to assume that all the unknowns in the expression have that *property*.

```
> simplify( expr, assume=real );
```

$$\mathrm{signum}(x)\,x\,\mathrm{signum}(y)\,y$$

```
> simplify( expr, assume=positive );
```

$$x\,y$$

You can also use the general assume facility to place assumptions on individual variables. See *The Assume Facility* on page 166.

## Simplification with Side Relations

Sometimes you can simplify an expression using your own special-purpose transformation rule. The `simplify` command allows you do to this by means of *side relations*.

```
> expr := x*y*z + x*y + x*z + y*z;
```

$$expr := x\,y\,z + x\,y + x\,z + y\,z$$

```
> simplify( expr, { x*z=1 } );
```

$$x\,y + y\,z + y + 1$$

You may give one or more side relations in a set or list. The `simplify` command uses the given equations as additional allowable simplifications.

Specifying the order in which `simplify` performs the simplification provides another level of control.

```
> expr := x^3 + y^3;
```

$$expr := x^3 + y^3$$

```
> siderel := x^2 + y^2 = 1;
```

$$siderel := x^2 + y^2 = 1$$

```
> simplify( expr, {siderel}, [x,y] );
```

$$y^3 - x\,y^2 + x$$

```
> simplify( expr, {siderel}, [y,x] );
```

$$x^3 - y\,x^2 + y$$

In the first case, Maple makes the substitution $x^2 = 1 - y^2$ into the expression, then attempts to make substitutions for $y^2$ terms; not finding any, it stops.

In the second case, Maple makes the substitution $y^2 = 1 - x^2$ into the expression, then attempts to make substitutions for $x^2$ terms; not finding any, it stops.

Gröbner Basis manipulations of polynomials are the basis of how `simplify` works. For more information about how this works, see the help page `?simplify,siderels`.

## Sorting Algebraic Expressions

Maple prints the terms of a polynomial in the order the polynomial was first created. You may want to sort the polynomial by decreasing degree. The `sort` command makes this possible.

```
> poly := 1 + x^4 - x^2 + x + x^3;
```

$$poly := 1 + x^4 - x^2 + x + x^3$$

```
> sort( poly );
```

$$x^4 + x^3 - x^2 + x + 1$$

Note that `sort` re-orders algebraic expressions in-place, replacing the original polynomial with the sorted copy.

```
> poly;
```

$$x^4 + x^3 - x^2 + x + 1$$

You can sort multivariate polynomials in two ways, by total degree or by lexicographic order. The default case is total degree, which sorts terms into descending order of degree. With this sort, if two terms have the same degree, it sorts those terms by lexicographic order (in other words, *a* comes before *b* and so forth).

```
> sort( x+x^3 + w^5 + y^2 + z^4, [w,x,y,z] );
```

$$w^5 + z^4 + x^3 + y^2 + x$$

```
> sort( x^3*y + y^2*x^2, [x,y] );
```

$$x^3 y + x^2 y^2$$

```
> sort( x^3*y + y^2*x^2 + x^4, [x,y] );
```

$$x^4 + x^3 y + x^2 y^2$$

Note that the order of the variables in the list determines the ordering of the expression.

```
> sort( x^3*y + y^2*x^2, [x,y] );
```

$$x^3 y + x^2 y^2$$

```
> sort( x^3*y + y^2*x^2, [y,x] );
```

$$y^2 x^2 + y x^3$$

You can also sort the entire expression by lexicographic ordering, using the `plex` option to the `sort` command.

```
> sort( x + x^3 + w^5 + y^2 + z^4, [w,x,y,z], plex );
```

$$w^5 + x^3 + x + y^2 + z^4$$

Again, the order in which you give the unknowns in the call to `sort` determines the ordering.

```
> sort( x + x^3 + w^5 + y^2 + z^4, [x,y,z,w], plex );
```

$$x^3 + x + y^2 + z^4 + w^5$$

The sort command can also sort lists. See *Sorting Lists* on page 174.

## Converting Between Equivalent Forms

You can write many mathematical functions, in several equivalent forms. For example, you can express $\sin(x)$ in terms of the exponential function. The convert command's capabilities include making these conversions.

```
> convert( sin(x), exp );
```

$$-\frac{1}{2} I \left( e^{(Ix)} - \frac{1}{e^{(Ix)}} \right)$$

```
> convert( cot(x), sincos );
```

$$\frac{\cos(x)}{\sin(x)}$$

```
> convert( arccos(x), ln );
```

$$-I \ln(x + I \sqrt{-x^2 + 1})$$

```
> convert( binomial(n,k), factorial );
```

$$\frac{n!}{k! (n - k)!}$$

The parfrac argument indicates partial fractions.

```
> convert( (x^5+1) / (x^4-x^2), parfrac, x );
```

$$x + \frac{1}{x - 1} - \frac{1}{x^2}$$

You can also use convert to find a fractional approximation to a floating-point number.

```
> convert( .3284879342, rational );
```

$$\frac{19615}{59713}$$

Note that conversions are not necessarily mutually inverse.

```
> convert( tan(x), exp );
```

$$-\frac{I ((e^{(Ix)})^2 - 1)}{(e^{(Ix)})^2 + 1}$$

```
> convert( %, trig );
```

$$-\frac{I\,((\cos(x) + I\,\sin(x))^2 - 1)}{(\cos(x) + I\,\sin(x))^2 + 1}$$

The `simplify` command reveals that this expression is $\tan(x)$.

```
> simplify( % );
```

$$\frac{\sin(x)}{\cos(x)}$$

You can also use the `convert` command to perform structural manipulations on Maple objects. See *Changing the Type of an Expression* on page 185.

## 5.2 The Assume Facility

The *assume facility* is a set of routines for dealing with properties of unknowns. assume helps Maple deal better with simplifications of symbolic expressions, especially with multiple-valued functions, for example, the square root.

```
> sqrt(a^2);
```

$$\sqrt{a^2}$$

Maple cannot simplify this, as the result is different for positive and negative values of $a$. Stating an assumption about the value of $a$ allows Maple to simplify the expression.

```
> assume( a>0 );
> sqrt(a^2);
```

$$\tilde{a}$$

The tilde (~) on a variable indicates that an assumption has been made about it. New assumptions replace old ones.

```
> assume( a<0 );
> sqrt(a^2);
```

$$-\tilde{a}$$

Use the about command to get information about the assumptions on an unknown.

```
> about(a);
```

```
Originally a, renamed a~:
  is assumed to be: RealRange(-infinity,Open(0))
```

Use the `additionally` command to make additional assumptions about unknowns.

```
> assume(m, nonneg);
> additionally( m<=0 );
> about(m);
```

```
Originally m, renamed m~:
  is assumed to be: 0
```

Many functions make use of the assumptions on an unknown. The `frac` command returns the fractional part of a number.

```
> frac(n);
```

$$\mathrm{frac}(n)$$

```
> assume(n, integer);
> frac(n);
```

$$0$$

The following limit depends on $b$.

```
> limit(b*x, x=infinity);
```

$$\mathrm{signum}(b)\,\infty$$

```
> assume( b>0 );
> limit(b*x, x=infinity);
```

$$\infty$$

You can use `infolevel` to have Maple report the details of what a command is doing.

```
> infolevel[int] := 2;
```

$$\mathit{infolevel}_{int} := 2$$

```
> int( exp(c*x), x=0..infinity );
```

```
Definite integration: Can't determine if the integral \
is convergent.
Need to know the sign of --> -c
Will now try indefinite integration and then take limi\
```

```
ts.
int/indef1:    first-stage indefinite integration
int/indef2:    second-stage indefinite integration
int/indef2:    applying derivative-divides
int/indef1:    first-stage indefinite integration
```

$$\lim_{x \to \infty} \frac{e^{(cx)} - 1}{c}$$

The int command needs to know the sign of c (or rather the sign of -c).

```
> assume( c>0 );
> int( exp(c*x), x=0..infinity );
```

```
int/cook/nogo1:   Given Integral      Int(exp(x),x = 0
.. infinity)
Fits into this pattern:
Int(exp(-Ucplex*x^S1-U2*x^S2)*x^N*ln(B*x^DL)^M*cos(C1*
x^R)/((A0+A1*x^D)^P),x = t1 .. t2)
int/cook/IIntd1:
--> U must be <= 0 for converging integral
--> will use limit to find if integral is +infinity
--> or - infinity or undefined
```

$$\infty$$

Logarithms are multiple-valued; for general complex values of $x$, $\ln(e^x)$ is different from $x$.

```
> ln( exp( 3*Pi*I ) );
```

$$I \pi$$

Therefore, Maple does not simplify the following unless it assumes $x$ is real.

```
> ln(exp(x));
```

$$\ln(e^x)$$

```
> assume(x, real);
> ln(exp(x));
```

$$x\tilde{}$$

You can use the is command to test properties of unknowns.

```
> is( c>0 );
```

$$true$$

```
> is(x, complex);
```

$$true$$

```
> is(x, real);
```

$$true$$

In this next example, Maple still assumes that the variable a is negative.

```
> eq := xi^2 = a;
```

$$eq := \xi^2 = a\tilde{}$$

```
> solve( eq, {xi} );
```

$$\{\xi = I\sqrt{-a\tilde{}}\}, \{\xi = -I\sqrt{-a\tilde{}}\}$$

To remove assumptions that you make on a name, simply unassign the name. However, the expression eq still refers to a˜.

```
> eq;
```

$$\xi^2 = a\tilde{}$$

*You must remove the assumption on* a *inside* eq *before* you remove the assumption on a. First, remove the assumptions on a inside eq.

```
> eq := subs( a='a', eq );
```

$$eq := \xi^2 = a$$

Then, unassign a.

```
> a := 'a';
```

$$a := a$$

See ?assume for more information.

## 5.3 Structural Manipulations

Structural manipulations include selecting and changing parts of an object, using knowledge of the structure or internal representation of an object rather than working with the expression as a purely mathematical expression. In the special cases of lists and sets, choosing an element is straightforward.

```
> L := { Z, Q, R, C, H, 0 };
```

$$L := \{O, R, H, C, Z, Q\}$$

> L[3];

$$H$$

Selecting elements from lists and sets is easy, which makes manipulating them straightforward. The concept of what constitutes the parts of a general expression is more difficult. However, many of the commands that manipulate lists and sets also apply to general expressions.

## Mapping a Function onto a List or Set

You may want to apply a function or command to each of the elements rather than to the object as a whole. This situation arises frequently with sets and lists. The map command allows you to do this.

> f( [a, b, c] );

$$f([a, b, c])$$

> map( f, [a, b, c] );

$$[f(a), f(b), f(c)]$$

> map( expand, { (x+1)*(x+2), x*(x+2) } );

$$\{x^2 + 3x + 2, x^2 + 2x\}$$

> map( x->x^2, [a, b, c] );

$$[a^2, b^2, c^2]$$

If you give map more than two arguments, it passes the extra arguments to the function.

> map( f, [a, b, c], p, q );

$$[f(a, p, q), f(b, p, q), f(c, p, q)]$$

> map( diff, [ (x+1)*(x+2), x*(x+2) ], x );

$$[2x + 3, 2x + 2]$$

The map2 command is closely related to map. Where map sequentially replaces the first argument of a function, the map2 command replaces the second argument to a function.

> map2( f, p, [a,b,c], q, r );

$$[f(p, a, q, r), f(p, b, q, r), f(p, c, q, r)]$$

You can use map2 to list all the partial derivatives of an expression.

```
> map2( diff, x^y/z, [x,y,z] );
```

$$\left[\frac{x^y\, y}{x\, z},\ \frac{x^y \ln(x)}{z},\ -\frac{x^y}{z^2}\right]$$

You can even use map2 in conjunction with map when applying them to subelements.

```
> map2( map, { [a,b], [c,d], [e,f] }, p, q );
```

$$\{[a(p,\,q),\,b(p,\,q)],\,[c(p,\,q),\,d(p,\,q)],\,[e(p,\,q),\,f(p,\,q)]\}$$

You may also use the seq command, to generate sequences resembling the output from map. Here seq generates a sequence by applying the function f to the elements of a set and a list.

```
> seq( f(i), i={a,b,c} );
```

$$f(c),\ f(b),\ f(a)$$

```
> seq( f(p, i, q, r), i=[a,b,c] );
```

$$f(p,\,a,\,q,\,r),\ f(p,\,b,\,q,\,r),\ f(p,\,c,\,q,\,r)$$

Here is Pascal's Triangle.

```
> L := [ seq( i, i=0..5 ) ];
```

$$L := [0,\, 1,\, 2,\, 3,\, 4,\, 5]$$

```
> [ seq( [ seq( binomial(n,m), m=L ) ], n=L ) ];
```

$$[[1,\, 0,\, 0,\, 0,\, 0,\, 0],\, [1,\, 1,\, 0,\, 0,\, 0,\, 0],\, [1,\, 2,\, 1,\, 0,\, 0,\, 0],$$
$$[1,\, 3,\, 3,\, 1,\, 0,\, 0],\, [1,\, 4,\, 6,\, 4,\, 1,\, 0],\, [1,\, 5,\, 10,\, 10,\, 5,\, 1]]$$

```
> map( print, % );
```

$$[1,\, 0,\, 0,\, 0,\, 0,\, 0]$$
$$[1,\, 1,\, 0,\, 0,\, 0,\, 0]$$
$$[1,\, 2,\, 1,\, 0,\, 0,\, 0]$$
$$[1,\, 3,\, 3,\, 1,\, 0,\, 0]$$
$$[1,\, 4,\, 6,\, 4,\, 1,\, 0]$$
$$[1,\, 5,\, 10,\, 10,\, 5,\, 1]$$
$$[]$$

The add and mul commands work like seq except that they generate sums and products, respectively, instead of sequences.

```
> add( i^2, i=[5, y, sin(x), -5] );
```

$$50 + y^2 + \sin(x)^2$$

The map, map2, seq, add, and mul commands can also act on general expressions. See *The Parts of an Expression* on page 176.

## Choosing Elements from a List or Set

You can select certain elements from a list or a set, if you have a boolean-valued function that determines which elements to select. The following boolean-valued function returns true if its argument is larger than three.

```
> large := x -> is(x > 3);
```

$$large := x \rightarrow is(3 < x)$$

You can now use the select command to choose the elements in a list or set that satisfy large.

```
> L := [ 8, 2.95, Pi, sin(9) ];
```

$$L := [8, 2.95, \pi, \sin(9)]$$

```
> select( large, L );
```

$$[8, \pi]$$

Similarly, the remove command removes the elements from L that satisfy large and displays as output the remaining elements.

```
> remove( large, L );
```

$$[2.95, \sin(9)]$$

You can use the type command to determine the type of an expression.

```
> type( 3, numeric );
```

$$true$$

```
> type( cos(1), numeric );
```

$$false$$

The syntax of select here passes the third argument, numeric, to the type command.

```
> select( type, L, numeric );
```

$$[8, 2.95]$$

See *The Parts of an Expression* on page 176 for more on types and how you can also use select and remove on a general expression.

## Merging Two Lists

Sometimes you need to merge two lists in some way. Here is a list of *x*-values and a list of *y*-values.

```
> X := [ seq( ithprime(i), i=1..6 ) ];
```

$$X := [2, 3, 5, 7, 11, 13]$$

```
> Y := [ seq( binomial(6, i), i=1..6 ) ];
```

$$Y := [6, 15, 20, 15, 6, 1]$$

To plot the *y*-values against the *x*-values, construct a list of lists: [ [*x1*,*y1*], [*x2*,*y2*], ... ]. That is, for each pair of values, construct a two-element list.

```
> pair := (x,y) -> [x, y];
```

$$pair := (x, y) \rightarrow [x, y]$$

Now the zip command can merge the lists X and Y according to the binary function pair.

```
> P := zip( pair, X, Y );
```

$$P := [[2, 6], [3, 15], [5, 20], [7, 15], [11, 6], [13, 1]]$$

```
> plot( P );
```

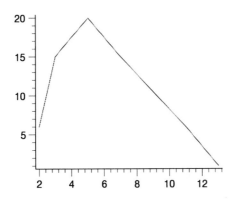

If the two lists have different length, then zip returns a list as long as the shorter one.

```
> zip( (x,y) -> x.y, [a,b,c,d,e,f], [1,2,3] );
```

$$[a1, b2, c3]$$

You may specify a fourth argument to zip. Then zip returns a list as long as the longer input list, using the fourth argument for the missing values.

```
> zip( (x,y) -> x.y, [a,b,c,d,e,f], [1,2,3], 99 );
```

$$[a1, b2, c3, d99, e99, f99]$$

```
> zip( igcd, [7657,342,876], [34,756,213,346,123], 6! );
```

$$[1, 18, 3, 2, 3]$$

The zip command can also merge vectors. See ?zip for more information.

## Sorting Lists

A list is a fundamental order-preserving data structure in Maple. The elements in a list stay in the same order as when the list was first created. You may want to create a copy of a list sorted in another order, using the sort command. The sort command sorts lists, among other things, in ascending order. It sorts a list of numbers in numerical order.

```
> sort( [1,3,2,4,5,3,6,3,6] );
```

$$[1, 2, 3, 3, 3, 4, 5, 6, 6]$$

The sort command also sorts a list of strings in lexicographic order.

```
> sort( ["Mary", "has", "a", "little", "lamb"] );
```

$$[\text{“Mary”, “a”, “has”, “lamb”, “little”}]$$

If you mix numbers and strings, or if the list contains expressions other than numbers or strings, sort uses the machine addresses, which are session dependent.

```
> sort( [x, 1, "apple"] );
```

$$[1, x, \text{“apple”}]$$

```
> sort( [-5, 10, sin(34)] );
```

$$[-5, 10, \sin(34)]$$

Note that to Maple, $\pi$ is not numeric.

```
> sort( [4.3, Pi, 2/3] );
```

$$\left[\pi, 4.3, \frac{2}{3}\right]$$

You can specify a boolean function to define an ordering for the list. The boolean function must take two arguments and return true if the first argument should come before the second. You can use this to sort a list of numbers in descending order.

```
> sort( [3.12, 1, 1/2], (x,y) -> evalb( x>y ) );
```

$$\left[ 3.12, \ 1, \ \frac{1}{2} \right]$$

The is command can compare constants like $\pi$ and $\sin(5)$ with pure numbers.

```
> bf := (x,y) -> is( x < y );
```

$$bf := (x, \ y) \rightarrow is(x < y)$$

```
> sort( [4.3, Pi, 2/3, sin(5)], bf );
```

$$\left[ \sin(5), \ \frac{2}{3}, \ \pi, \ 4.3 \right]$$

You can also sort strings by length.

```
> shorter := (x,y) -> evalb( length(x) < length(y) );
```

$$shorter := (x, \ y) \rightarrow evalb( \ length(x) < length(y))$$

```
> sort( ["Mary", "has", "a", "little", "lamb"], shorter );
```

$$[\text{``a''}, \ \text{``has''}, \ \text{``lamb''}, \ \text{``Mary''}, \ \text{``little''}]$$

Maple does not have a built-in method for sorting lists of mixed strings and numbers, other than by machine address. To sort a mixed list of strings and numbers, you can do the following.

```
> big_list := [1,"d",3,5,2,"a","c","b",9];
```

$$big\_list := [1, \ \text{``d''}, \ 3, \ 5, \ 2, \ \text{``a''}, \ \text{``c''}, \ \text{``b''}, \ 9]$$

Make two lists from the original, one consisting of numbers and one consisting of strings.

```
> list1 := select( type, big_list, string );
```

$$list1 := [\text{``d''}, \ \text{``a''}, \ \text{``c''}, \ \text{``b''}]$$

```
> list2 := select( type, big_list, numeric );
```

$$list2 := [1, \ 3, \ 5, \ 2, \ 9]$$

Then sort the two lists independently.

```
> list1 := sort(list1);
```

$$list1 := [\text{``a''}, \text{``b''}, \text{``c''}, \text{``d''}]$$

```
> list2 := sort(list2);
```

$$list2 := [1, 2, 3, 5, 9]$$

Finally, stack the two lists together.

```
> sorted_list := [ op(list1), op(list2) ];
```

$$sorted\_list := [\text{``a''}, \text{``b''}, \text{``c''}, \text{``d''}, 1, 2, 3, 5, 9]$$

The sort command can also sort algebraic expressions. See *Sorting Algebraic Expressions* on page 163.

*Choosing Elements from a List or Set* on page 172 gives more information about the commands in this example.

## The Parts of an Expression

In order to manipulate the details of an expression you must select the individual parts. Three easy cases for doing this involve equations, ranges, and fractions. The lhs command selects the left-hand side of an equation.

```
> eq := a^2 + b^ 2 = c^2;
```

$$eq := a^2 + b^2 = c^2$$

```
> lhs( eq );
```

$$a^2 + b^2$$

The rhs command similarly selects the right-hand side.

```
> rhs( eq );
```

$$c^2$$

The lhs and rhs commands also work for ranges.

```
> lhs( 2..5 );
```

$$2$$

```
> rhs( 2..5 );
```

$$5$$

```
> eq := x = -2..infinity;
```

$$eq := x = -2..\infty$$

```
> lhs( eq );
```

$$x$$

```
> rhs( eq );
```

$$-2..\infty$$

```
> lhs( rhs(eq) );
```

$$-2$$

```
> rhs( rhs(eq) );
```

$$\infty$$

The numer and denom commands extract the numerator and denominator, respectively, from a fraction.

```
> numer( 2/3 );
```

$$2$$

```
> denom( 2/3 );
```

$$3$$

```
> fract := ( 1+sin(x)^3-y/x) / ( y^2 - 1 + x );
```

$$fract := \frac{1 + \sin(x)^3 - \frac{y}{x}}{y^2 - 1 + x}$$

```
> numer( fract );
```

$$x + \sin(x)^3 x - y$$

```
> denom( fract );
```

$$x\,(y^2 - 1 + x)$$

Consider the expression

```
> expr := 3 + sin(x) + 2*cos(x)^2*sin(x);
```

$$expr := 3 + \sin(x) + 2\cos(x)^2 \sin(x)$$

The whattype command identifies expr as a sum.

```
> whattype( expr );
```

$$`+`$$

Use the op command to list the terms of a sum or, in general, the operands of an expression.

```
> op( expr );
```

$$3, \sin(x), 2\cos(x)^2 \sin(x)$$

expr consists of three terms. Use the nops command to count the number of operands in an expression.

```
> nops( expr );
```

$$3$$

Since op(expr) is a sequence, you can pick out, for example, the third term as follows.

```
> term3 := op(expr)[3];
```

$$term3 := 2\cos(x)^2 \sin(x)$$

term3 is a product of three factors.

```
> whattype( term3 );
```

$$`*`$$

```
> nops( term3 );
```

$$3$$

```
> op( term3 );
```

$$2, \cos(x)^2, \sin(x)$$

Retrieve the second factor in term3 in the following manner.

```
> factor2 := op(term3)[2];
```

$$factor2 := \cos(x)^2$$

It is an exponentiation.

```
> whattype( factor2 );
```

$$`^`$$

factor2 has two operands.

```
> op( factor2 );
```

$$\cos(x), 2$$

The first operand is a function and has just one operand.

```
> op1 := op(factor2)[1];
```

$$op1 := \cos(x)$$

```
> whattype( op1 );
```

*function*

```
> op( op1 );
```

*x*

x is a symbol.

```
> whattype( op(op1) );
```

*symbol*

Since you did not assign a value to *x*, it has only one operand, namely itself.

```
> nops( op(op1) );
```

1

```
> op( op(op1) );
```

*x*

You can represent the result of finding the operands of the operands of an expression as a picture called an *expression tree*. The expression tree for expr looks like this.

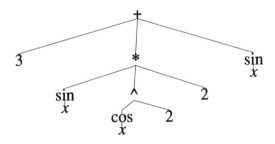

The operands of a list or set are the elements.

```
> op( [a,b,c] );
```

*a, b, c*

```
> op( {d,e,f} );
```

*f, e, d*

*Mapping a Function onto a List or Set* on page 170 describes how the map command applies a function to all the elements of a list or set. The functionality of map extends to general expressions.

```
> map( f, x^2 );
```

$$f(x)^{f(2)}$$

The select and remove commands, described in *Choosing Elements from a List or Set* on page 172, also work on general expressions.

```
> large := z -> evalb( is(z>3) = true );
```

$$large := z \rightarrow evalb(is(3 < z) = true)$$

```
> remove( large, 5+8*sin(x) - exp(9) );
```

$$8\sin(x) - e^9$$

Maple has a number of commands that are useful as the boolean function in a call to select or remove. The has command determines whether an expression contains a certain subexpression.

```
> has( x*exp(cos(t^2)), t^2 );
```

$$true$$

```
> has( x*exp(cos(t^2)), cos );
```

$$true$$

Some of the solutions to the following set of equations contain RootOf's.

```
> sol := { solve( { x^2*y^2 = b*y, x^2-y^2 = a*x },
>                  {x,y} ) };
```

$$sol := \{\{y = 0, x = 0\}, \{y = 0, x = a\}, \{$$
$$x = RootOf(\_Z^6 - b^2 - a \_Z^5),$$
$$y = \frac{b}{RootOf(\_Z^6 - b^2 - a \_Z^5)^2}\}\}$$

You can use select and has to pick those solutions out.

```
> select( has, sol, RootOf );
```

$$\{\{x = RootOf(\_Z^6 - b^2 - a \_Z^5),$$
$$y = \frac{b}{RootOf(\_Z^6 - b^2 - a \_Z^5)^2}\}\}$$

You can also select or remove subexpressions by type. The type command determines if an expression is of a certain type.

```
> type( 3+x, '+' );
```

$$true$$

Here the select command passes its third argument, '+', to type.

```
> expr := ( 3+x ) * x^2 * sin( 1+sqrt(Pi) );
```

$$expr := (3 + x)\, x^2 \sin(1 + \sqrt{\pi})$$

```
> select( type, expr, '+' );
```

$$3 + x$$

The hastype command determines if an expression contains a subexpression of a certain type.

```
> hastype( sin( 1+sqrt(Pi) ), '+' );
```

$$true$$

You can use the combination select(hastype,...) to select the operands of an expression that contain a certain type.

```
> select( hastype, expr, '+' );
```

$$(3 + x) \sin(1 + \sqrt{\pi})$$

If you are interested in the subexpressions of a certain type rather than the operands that contain them, use the indets command.

```
> indets( expr, '+' );
```

$$\{1 + \sqrt{\pi},\ 3 + x\}$$

The two RootOf's in sol above are of type RootOf. Since the two RootOf's are identical, the set that indets returns contains only one element.

```
> indets( sol, RootOf );
```

$$\{\text{RootOf}(\_Z^6 - b^2 - a\,\_Z^5)\}$$

Not all commands are their own type, as is RootOf, but you can use the structured type specfunc(*type*, *name*). This type matches the function *name* with arguments of type *type*.

```
> type( diff(y(x), x), specfunc(anything, diff) );
```

$$true$$

You can use this to find all the derivatives in a large differential equation.

```
> DE := expand( diff( cos(y(t)+t)*sin(t*z(t)), t ) )
>      + diff(x(t), t);
```

$$DE := -\sin(t\,z(t))\,\sin(y(t))\,\cos(t)\,\left(\frac{\partial}{\partial t}\,y(t)\right)$$

$$-\sin(t\,z(t))\,\sin(y(t))\,\cos(t)$$

$$-\sin(t\,z(t))\,\cos(y(t))\,\sin(t)\,\left(\frac{\partial}{\partial t}\,y(t)\right)$$

$$-\sin(t\,z(t))\,\cos(y(t))\,\sin(t)$$

$$+\cos(t\,z(t))\,\cos(y(t))\,\cos(t)\,z(t)$$

$$+\cos(t\,z(t))\,\cos(y(t))\,\cos(t)\,t\,\left(\frac{\partial}{\partial t}\,z(t)\right)$$

$$-\cos(t\,z(t))\,\sin(y(t))\,\sin(t)\,z(t)$$

$$-\cos(t\,z(t))\,\sin(y(t))\,\sin(t)\,t\,\left(\frac{\partial}{\partial t}\,z(t)\right)+\left(\frac{\partial}{\partial t}\,x(t)\right)$$

```
> indets( DE, specfunc(anything, diff) );
```

$$\left\{\frac{\partial}{\partial t}\,x(t),\ \frac{\partial}{\partial t}\,y(t),\ \frac{\partial}{\partial t}\,z(t)\right\}$$

The following operands of DE contain the derivatives.

```
> select( hastype, DE, specfunc(anything, diff) );
```

$$-\sin(t\,z(t))\,\sin(\,y(t))\,\cos(t)\,\left(\frac{\partial}{\partial t}\,y(t)\right)$$

$$-\sin(t\,z(t))\,\cos(y(t))\,\sin(t)\,\left(\frac{\partial}{\partial t}\,y(t)\right)$$

$$+\cos(t\,z(t))\,\cos(y(t))\,\cos(t)\,t\,\left(\frac{\partial}{\partial t}\,z(t)\right)$$

$$-\cos(t\,z(t))\,\sin(y(t))\,\sin(t)\,t\,\left(\frac{\partial}{\partial t}\,z(t)\right)+\left(\frac{\partial}{\partial t}\,x(t)\right)$$

DE has only one operand that is itself a derivative.

```
> select( type, DE, specfunc(anything, diff) );
```

$$\frac{\partial}{\partial t}\,x(t)$$

Maple recognizes many types. See `?type` for a partial list, and `?type, structured` for more on structured types, such as `specfunc`.

## Substitution

Often you want to substitute a value for a variable (i.e., evaluate an expression at a point). For example, if you need to solve the problem, "If $f(x) = \ln(\sin(x e^{\cos(x)}))$, find $f'(2)$," then you must substitute the value 2 for $x$ in the derivative. The `diff` command finds the derivative.

```
> y := ln( sin( x * exp(cos(x)) ) );
```

$$y := \ln(\sin(x \, e^{\cos(x)}))$$

```
> yprime := diff( y, x );
```

$$yprime := \frac{\cos(x \, e^{\cos(x)}) \, (e^{\cos(x)} - x \sin(x) \, e^{\cos(x)})}{\sin(x \, e^{\cos(x)})}$$

Now use the `eval` command to substitute a value for x in yprime.

```
> eval( yprime, x=2 );
```

$$\frac{\cos(2 \, e^{\cos(2)}) \, (e^{\cos(2)} - 2 \sin(2) \, e^{\cos(2)})}{\sin(2 \, e^{\cos(2)})}$$

The `evalf` command returns a floating-point approximation of the result.

```
> evalf( % );
```

$$-.1388047428$$

The subs command makes syntactical substitutions, not mathematical substitutions. This means that you can make substitutions for any subexpression.

```
> subs( cos(x)=3, yprime );
```

$$\frac{\cos(x \, e^3) \, (e^3 - x \sin(x) \, e^3)}{\sin(x \, e^3)}$$

But you are limited to subexpressions as Maple sees them.

```
> expr := a * b * c * a^b;
```

$$expr := a \, b \, c \, a^b$$

```
> subs( a*b=3, expr );
```

$$a \, b \, c \, a^b$$

To Maple, expr is a product of four factors.

```
> op( expr );
```

$$a, \ b, \ c, \ a^b$$

The product a*b is not a factor in expr. You can make the substitution a*b=3 in three ways: solve the subexpression for one of the variables

```
> subs( a=3/b, expr );
```

$$3\,c\,\left(\frac{3}{b}\right)^b$$

or use a side relation to simplify,

```
> simplify( expr, { a*b=3 } );
```

$$3\,c\,a^b$$

or use the algsubs command, which performs algebraic substitutions.

```
> algsubs( a*b=3, expr);
```

$$3\,c\,a^b$$

Note that in the first case all occurrences of a have been replaced by 3/b. Whereas, in the second and third cases both variables a and b remain in the result. Using simplify is usually preferred.

You can make several substitutions with one call to subs.

```
> expr := z * sin( x^2 ) + w;
```

$$expr := z\sin(x^2) + w$$

```
> subs( x=sqrt(z), w=Pi, expr );
```

$$z\sin(z) + \pi$$

The subs command makes the substitutions from left to right.

```
> subs( z=x, x=sqrt(z), expr );
```

$$\sqrt{z}\sin(z) + w$$

If you give a set or list of substitutions, subs makes those substitutions simultaneously.

```
> subs( { x=sqrt(Pi), z=3 }, expr );
```

$$3\sin(\pi) + w$$

Note that in general you must explicitly evaluate the result of a call to subs.

```
> eval( % );
```

$$\sqrt{z}\,\sin(z) + w$$

If you know you want the result fully evaluated, use the two-parameter version of `eval`.

Use the `subsop` command to substitute for a specific operand of an expression.

```
> expr := 5^x;
```

$$expr := 5^x$$

```
> op( expr );
```

$$5, x$$

```
> subsop( 1=t, expr );
```

$$t^x$$

The zeroth operand of a function is typically the name of the function.

```
> expr := cos(x);
```

$$expr := \cos(x)$$

```
> subsop( 0=sin, expr );
```

$$\sin(x)$$

*The Parts of an Expression* on page 176 explains the operands of an expression.

## Changing the Type of an Expression

You may find it necessary to convert an expression to another type. Here is the Taylor series for sin(x).

```
> f := sin(x);
```

$$f := \sin(x)$$

```
> t := taylor( f, x=0 );
```

$$t := x - \frac{1}{6}x^3 + \frac{1}{120}x^5 + O(x^6)$$

For example, you cannot plot a series, you must use `convert(..., polynom)` to convert it into a polynomial approximation first.

```
> p := convert( t, polynom );
```

$$p := x - \frac{1}{6}x^3 + \frac{1}{120}x^5$$

Similarly, the title of a plot must be a string, not a general expression. You can use convert(..., string) to convert an expression to a string.

```
> p_txt := convert( p, string );
```

$$p\_txt := \text{``x-1/6*x^3+1/120*x\^{} 5"}$$

```
> plot( p, x=-4..4, title=p_txt );
```

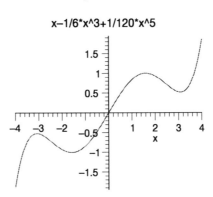

The cat command concatenates all its arguments to create a new string.

```
> ttl := cat( convert( f, string ),
>             " and its Taylor approximation ",
>             p_txt );
```

$$ttl := \text{``sin(x) and its Taylor\ approximation x-1/6*\\}$$

$$x\^{}3+1/120*x\^{}5"$$

```
> plot( [f, p], x=-4..4, title=ttl );
```

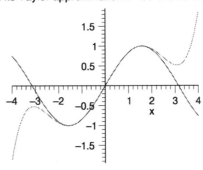

You can also convert a list to a set or a set to a list.

```
> L := [1,2,5,2,1];
```

$$L := [1, 2, 5, 2, 1]$$

```
> S := convert( L, set );
```

$$S := \{1, 2, 5\}$$

```
> convert( S, list );
```

$$[1, 2, 5]$$

The convert command can perform many other structural and mathematical conversions. See ?convert for more information.

## 5.4   Evaluation Rules

In a symbolic mathematics program such as Maple you encounter the issue of *evaluation*. If you assign the value y to x, the value z to y, and the value 5 to z, then to what should x evaluate?

### Levels of Evaluation

Maple, in most cases, does full evaluation of names. That is, when you use a name or symbol, Maple checks if the name or symbol has an assigned value. If it has a value, Maple substitutes the value for the name. If this value itself has an assigned value, Maple performs a substitution again, and so on, recursively, until no more substitutions are possible.

```
> x := y;
```

$$x := y$$

```
> y := z;
```

$$y := z$$

```
> z := 5;
```

$$z := 5$$

Now Maple evaluates x fully. That is, Maple substitutes y for x, z for y, and finally, 5 for z.

```
> x;
```

$$5$$

You can use the eval command to control the level of evaluation of an expression. If you call eval with just one argument, then eval evaluates that argument fully.

```
> eval(x);
```

$$5$$

A second argument to eval specifies how far you want to evaluate the first argument.

```
> eval(x, 1);
```

$$y$$

```
> eval(x, 2);
```

$$z$$

```
> eval(x, 3);
```

$$5$$

The main exceptions to the rule of full evaluation are special data structures like tables, matrices, and procedures, and the behavior of local variables inside a procedure.

## Last-Name Evaluation

The data structures array, table, matrix, and proc have a special evaluation behavior called *last-name evaluation.*

```
> x := y;
```

$$x := y$$

```
> y := z;
```

$$y := z$$

```
> z := array( [ [1,2], [3,4] ] );
```

$$z := \begin{bmatrix} 1 & 2 \\ 3 & 4 \end{bmatrix}$$

Maple substitutes y for x and z for y but it stops there since evaluation of the last name, z, would produce an array, one of the four special structures.

```
> x;
```

$$z$$

Maple uses last-name evaluation for arrays, tables, matrices, and procedures in order to retain compact representations of unassigned table entries (for example, T[3]) and unevaluated commands (for example, sin(x)). You can force full evaluation by calling eval explicitly.

```
> eval(x);
```

$$\begin{bmatrix} 1 & 2 \\ 3 & 4 \end{bmatrix}$$

```
> add2 := proc(x,y) x+y; end;
```

$$add2 := \mathbf{proc}(x,\ y)\ x + y\ \mathbf{end}$$

```
> add2;
```

$$add2$$

You can easily force full evaluation, using `eval` or `print`.

```
> eval(add2);
```

$$\mathbf{proc}(x,\ y)\ x + y\ \mathbf{end}$$

Note that full evaluation of Maple library procedures, by default, suppresses the code in the procedure. To illustrate this, examine the `erfi` command

```
> erfi;
```

$$erfi$$

```
> eval(erfi);
```

$$\mathbf{proc}(x::algebraic)\ \ldots\ \mathbf{end}$$

Set the `interface` variable `verboseproc` to 2, and then try again.

```
> interface( verboseproc=2 );
> eval(erfi);
```

$\mathbf{proc}(x::algebraic)$

$\quad \mathbf{option}\,`Copyright\ (c)\ 1996\ Waterloo\ Maple\ Inc.``$

$\quad All\ rights\ reserved.`;$

$\quad\quad \mathbf{if}\ nargs \neq 1\ \mathbf{then}$

$\quad\quad\quad ERROR(`expecting\ 1\ argument,\ got\ `.nargs)$

$\quad\quad \mathbf{elif}\ type(x,\ 'complex(float)')\ \mathbf{then}\ evalf(`erfi`(x))$

$\quad\quad \mathbf{elif}\ type(x,\ `*`)\ \mathbf{and}\ member(I,\ \{op(x)\})\ \mathbf{then}$

$\quad\quad\quad I \times erf(-I \times x)$

$\quad\quad \mathbf{elif}\ type(x,\ 'complex(numeric)')\ \mathbf{and}\ csgn(x) < 0\ \mathbf{then}$

$\quad\quad\quad - erfi(-x)$

$$\text{eliftype}(x, \text{`*`}) \text{ and type}(\text{op}(1, x), \text{'complex}(\textit{numeric})\text{')}$$

$$\text{and csgn}(\text{op}(1, x)) < 0 \textbf{then} - \text{erfi}(-x)$$

$$\textbf{elif } \text{type}(x, \text{`+`}) \text{ and traperror}(\text{sign}(x)) = -1 \textbf{ then}$$

$$- \text{erfi}(-x)$$

$$\textbf{else } \text{'erfi'}(x)$$

$$\textbf{fi}$$

**end**

The default value of verboseproc is 1.

```
> interface( verboseproc=1 );
```

The help page ?interface explains the possible settings of verboseproc and the other interface variables.

## One-Level Evaluation

Local variables of a procedure use one-level evaluation. That is, if you assign a local variable, then the result of evaluation is the value most recently assigned directly to that variable.

```
> test:=proc()
>     local x, y, z;
>     x := y;
>     y := z;
>     z := 5;
>     x;
> end:
> test();
```

$$y$$

Compare this evaluation with the similar interactive example in *Levels of Evaluation* on page 187. Full evaluation within a procedure is rarely necessary and can lead to inefficiency. If you require full evaluation within a procedure, use eval.

## Commands with Special Evaluation Rules

**The assigned and evaln Commands**   The functions assigned and evaln evaluate their arguments only to the point where the arguments first become names.

```
> x := y;
```

$$x := y$$

```
> y := z;
```

$$y := z$$

```
> evaln(x);
```

$$x$$

The `assigned` command checks if a name has a value assigned to it.

```
> assigned( x );
```

$$true$$

**The `seq` Command**  The seq command for creating expression sequences does not evaluate its arguments, so that even if a variable has an assigned value, seq can still use it as a counting variable.

```
> i := 2;
```

$$i := 2$$

```
> seq( i^2, i=1..5 );
```

$$1, 4, 9, 16, 25$$

```
> i;
```

$$2$$

Contrast this with the behavior of sum.

```
> sum( i^2, i=1..5 );
```

```
Error, (in sum) summation variable previously assigned
,               second argument evaluates to, 2 = 1
.. 5
```

You can easily solve this problem using single quotes.

## Quotation and Unevaluation

The Maple language supports the use of quotes to delay evaluation one level. Surrounding a name in single quotes (') prevents Maple from evaluating the name.

```
> i := 4;
```

$$i := 4$$

```
> i;
```

$$4$$

```
> 'i';
```

$$i$$

Use this method to avoid the following problem.

```
> i;
```

$$4$$

```
> sum( i^2, i=1..5 );
```

```
Error, (in sum) summation variable previously assigned
,                 second argument evaluates to, 4 = 1
.. 5
```

```
> sum( 'i^2', 'i'=1..5 );
```

$$55$$

```
> i;
```

$$4$$

Full evaluation of a quoted expression strips off one level of quotes.

```
> x := 0;
```

$$x := 0$$

```
> '''x'+1'';
```

$$"x' + 1'$$

```
> %;
```

$$'x' + 1$$

```
> %;
```

$$x + 1$$

```
> %;
```

$$1$$

Quoting an expression delays evaluation, but does not prevent automatic simplifications and arithmetic.

```
> '1-1';
```

$$0$$

```
> 'p+q-i-p+3*q';
```

$$4\,q - i$$

If you enclose a simple variable in single quotes, the result is the name of the variable. You may use this method to unassign a variable.

```
> x := 1;
```

$$x := 1$$

```
> x;
```

$$1$$

```
> x := 'x';
```

$$x := x$$

```
> x;
```

$$x$$

However, in general, you must use evaln.

```
> i := 4;
```

$$i := 4$$

```
> a[i] := 9;
```

$$a_4 := 9$$

Note that 'a[i]' is a[i] not a[4].

```
> 'a[i]';
```

$$a_i$$

You must use evaln to unassign a[i].

```
> evaln( a[i] );
```

$$a_4$$

```
> a[i] := evaln( a[i] );
```

$$a_4 := a_4$$

## Using Quoted Variables as Function Arguments

Some Maple commands use names as a way to pass back information in addition to the standard return value.

```
> divide( x^2-1, x-1, 'q' );
```

$$true$$

This divide command assigns the quotient to the global name, q.

```
> q;
```

$$x + 1$$

Remember to use a quoted name to ensure that you are not passing a variable with an assigned value into the procedure. You can avoid the need for quotes if you ensure that the name you use has no previously assigned value.

```
> q := 2;
```

$$q := 2$$

```
> divide( x^2-y^2, x-y, q );
```

```
Error, wrong number (or type) of parameters in functio\
n divide
```

```
> q := evaln(q);
```

$$q := q$$

```
> divide( x^2-y^2, x-y, q );
```

$$true$$

```
> q;
```

$$y + x$$

The rem, quo, irem, and iquo commands behave in a similar manner.

## Concatenation of Names

Concatenation is a way to form new variable names based on others.

```
> a.b;
```

$$ab$$

The concatenation operator, ".", in a name causes evaluation of the right-hand side of the operator, but not the left.

```
> a := x;
```

$$a := x$$

```
> b := 2;
```

$$b := 2$$

```
> a.b;
```

$$a2$$

```
> c := 3;
```

$$c := 3$$

```
> a.b.c;
```

$$a23$$

If a name does not evaluate to a single symbol, Maple does not evaluate a concatenation.

```
> a := x;
```

$$a := x$$

```
> b := y+1;
```

$$b := y + 1$$

```
> new_name := a.b;
```

$$new\_name := a.(y + 1)$$

```
> y := 3;
```

$$y := 3$$

```
> new_name;
```

$$a4$$

You can use concatenated names to assign, and to create expressions.

```
> i := 1;
```

$$i := 1$$

```
> b.i := 0;
```

$$b1 := 0$$

You need single quotes here.

```
> sum( 'a.k' * x^k, k=0..8 );
```

$$a0 + a1\,x + a2\,x^2 + a3\,x^3 + a4\,x^4 + a5\,x^5 + a6\,x^6 + a7\,x^7$$
$$+ a8\,x^8$$

If you leave out the single quotes, Maple evaluates a.k to ak.

```
> sum( a.k * x^k, k=0..8 );
```

$$ak + ak\,x + ak\,x^2 + ak\,x^3 + ak\,x^4 + ak\,x^5 + ak\,x^6 + ak\,x^7$$
$$+ ak\,x^8$$

You can also use concatenation to form title strings for plots.

## 5.5  Conclusion

In this chapter, you have seen how to perform many kinds of manipulations of expressions, from adding two equations to selecting individual parts of a general expression. In general, no rule specifies which form of an expression is the simplest. But, the commands you have seen in this chapter allow you to turn an expression into many forms, often the ones *you* would consider simplest. If not, you can use side relations to specify your own simplification rules, or assumptions to tell Maple that certain unknowns have certain properties.

You have also seen that Maple in most cases uses full evaluation of variables. Some exceptions exist which include last-name evaluation for certain large data structures, one-level evaluation for local variables in a procedure, and delayed evaluation with single quotes.

# Examples from Calculus

This chapter provides examples of how Maple can help you present and solve problems from calculus. The first section describes elementary concepts such as the derivative and the integral, the second section treats ordinary differential equations in some depth, and the third section concerns partial differential equations.

## 6.1 Introductory Calculus

This section contains a number of examples of how to illustrate ideas and solve problems from calculus. The student package contains many commands that are especially useful in this area.

### The Derivative

This section illustrates the graphical meaning of the derivative: the slope of the tangent line. Then it shows you how to find the set of inflection points for a function.

Define the function $f: x \mapsto \exp(\sin(x))$ in the following manner.

```
> f := x -> exp( sin(x) );
```

$$f := x \to e^{\sin(x)}$$

Find the derivative of $f$ evaluated at $x_0 = 1$.

```
> x0 := 1;
```

$$x0 := 1$$

$p_0$ and $p_1$ are two points on the graph of $f$.

```
> p0 := [ x0, f(x0) ];
```

$$p0 := [1,\, e^{\sin(1)}]$$

```
> p1 := [ x0+h, f(x0+h) ];
```

$$p1 := [1+h,\, e^{\sin(1+h)}]$$

The slope command from the student package can find the slope of the secant line through $p_0$ and $p_1$.

```
> with(student):
> m := slope( p0, p1 );
```

$$m := -\frac{e^{\sin(1)} - e^{\,\sin(1+h)}}{h}$$

If $h = 1$, the slope is

```
> eval( m, h=1 );
```

$$-e^{\sin(1)} + e^{\sin(2)}$$

The evalf command gives a floating-point approximation.

```
> evalf( % );
```

$$.162800903$$

When $h$ tends to zero, the slope seems to converge.

```
> h_values := [ seq( 1/i^2, i=1..20 ) ];
```

$$h\_values := [1,\, \frac{1}{4},\, \frac{1}{9},\, \frac{1}{16},\, \frac{1}{25},\, \frac{1}{36},\, \frac{1}{49},\, \frac{1}{64},\, \frac{1}{81},\, \frac{1}{100},\, \frac{1}{121},\, \frac{1}{144},$$
$$\frac{1}{169},\, \frac{1}{196},\, \frac{1}{225},\, \frac{1}{256},\, \frac{1}{289},\, \frac{1}{324},\, \frac{1}{361},\, \frac{1}{400}]$$

```
> seq( evalf(m), h=h_values );
```

$$.162800903,\ 1.053234750,\ 1.17430578,\ 1.21091762,$$
$$1.22680697,\ 1.23515485,\ 1.2400915,\ 1.2432565,$$
$$1.2454086,\ 1.2469391,\ 1.2480669,\ 1.2489216,$$
$$1.2495855,\ 1.2501111,\ 1.2505343,\ 1.2508805,$$
$$1.2511671,\ 1.2514069,\ 1.2516098,\ 1.2517828$$

The following is the equation of the secant line.

```
> y - p0[2] = m * ( x - p0[1] );
```

$$y - e^{\sin(1)} = -\frac{(e^{\sin(1)} - e^{\sin(1+h)})\,(x - 1)}{h}$$

The `isolate` command turns the equation into slope–intercept form.

```
> isolate( %, y );
```

$$y = -\frac{(e^{\sin(1)} - e^{\sin(1+h)})\,(x - 1)}{h} + e^{\sin(1)}$$

You need to turn the equation into a function.

```
> secant := unapply( rhs(%), x );
```

$$secant := x \rightarrow -\frac{(e^{\sin(1)} - e^{\sin(1+h)})\,(x - 1)}{h} + e^{\sin(1)}$$

You can now plot the secant and the function itself together for different values of $h$. First, make a sequence of plots. You can use the slope of the secant as the title.

```
> S := seq( plot( [f(x), secant(x)], x=0..4,
>                 view=[0..4, 0..4],
>                 title=convert(evalf(m), string) ),
>         h=h_values ):
```

The `display` command from the `plots` package can display the plots in sequence—that is, as an animation.

```
> with(plots):
```

```
> display( S, insequence=true, view=[0..4, 0..4] );
```

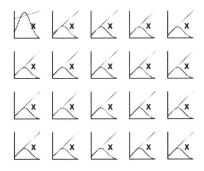

In the limit as $h$ tends to zero, the slope is

```
> Limit( m, h=0 );
```

$$\lim_{h \to 0} -\frac{e^{\sin(1)} - e^{\sin(1+h)}}{h}$$

The value of this limit is

```
> value( % );
```

$$e^{\sin(1)} \cos(1)$$

This answer is, of course, the value of $f'(x0)$. To see this, first take the derivative of $f$.

```
> diff( f(x), x );
```

$$\cos(x) e^{\sin(x)}$$

Then define the function $f1$ to be the first derivative of $f$.

```
> f1 := unapply( %, x );
```

$$f1 := x \to \cos(x) e^{\sin(x)}$$

Now you can see that $f1(x0)$ equals the limit above.

```
> f1(x0);
```

$$e^{\sin(1)} \cos(1)$$

If you can take the derivative once, you can take it twice.

```
> diff( f(x), x, x );
```

$$-\sin(x) e^{\sin(x)} + \cos(x)^2 e^{\sin(x)}$$

Again, define the function $f2$ to be the second derivative of $f$.

```
> f2 := unapply( %, x );
```

$$f2 := x \to -\sin(x) e^{\sin(x)} + \cos(x)^2 e^{\sin(x)}$$

When you plot $f$ and its first and second derivatives, you can see that $f$ is increasing whenever $f1$ is positive, and that $f$ is concave down whenever $f2$ is negative.

```
> plot( [f(x), f1(x), f2(x)], x=0..10 );
```

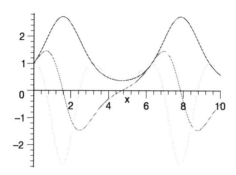

The graph of $f$ has an inflection point where the double derivative changes sign, and the double derivative may change sign at the values of $x$ where $f2(x)$ is zero.

```
> sol := { solve( f2(x)=0, x ) };
```

$$sol := \left\{ - \arctan\left(2\,\frac{-\dfrac{1}{2}+\dfrac{1}{2}\sqrt{5}}{\sqrt{-2+2\sqrt{5}}}\right) + \pi,\ \arctan\left(2\,\frac{-\dfrac{1}{2}+\dfrac{1}{2}\sqrt{5}}{\sqrt{-2+2\sqrt{5}}}\right),\right.$$

$$\arctan\left(-\frac{1}{2}-\frac{1}{2}\sqrt{5},\ \frac{1}{2}\sqrt{-2-2\sqrt{5}}\right),$$

$$\left.\arctan\left(-\frac{1}{2}-\frac{1}{2}\sqrt{5},\ -\frac{1}{2}\sqrt{-2-2\sqrt{5}}\right)\right\}$$

Two of these solutions are complex.

```
> evalf( sol );
```

$$\{.6662394325,\ 2.475353222,$$

$$-1.570796327 + 1.061275062\,I,$$

$$-1.570796327 - 1.061275062\,I\}$$

In this example, only the real solutions are of interest. You can use the select command to select the real constants from the solution set.

```
> infl := select( type, sol, realcons );
```

$$infl := \left\{ -\arctan\left(2\,\frac{-\frac{1}{2}+\frac{1}{2}\sqrt{5}}{\sqrt{-2+2\sqrt{5}}}\right) + \pi, \; \arctan\left(2\,\frac{-\frac{1}{2}+\frac{1}{2}\sqrt{5}}{\sqrt{-2+2\sqrt{5}}}\right) \right\}$$

```
> evalf( infl );
```

$$\{.6662394325, \; 2.475353222\}$$

You can see from the graph above that $f2$ actually does change signs at these $x$-values. Then, the set of inflection points is

```
> { seq( [x, f(x)], x=infl ) };
```

$$\left\{ \left[ -\arctan\left(2\,\frac{-\frac{1}{2}+\frac{1}{2}\sqrt{5}}{\sqrt{-2+2\sqrt{5}}}\right) + \pi, \right. \right.$$

$$e^{\left(2\,\frac{-1/2+1/2\sqrt{5}}{\sqrt{-2+2\sqrt{5}}\sqrt{1+4\frac{(-1/2+1/2\sqrt{5})^2}{-2+2\sqrt{5}}}}\right)} \Bigg],$$

$$\left[ \arctan\left(2\,\frac{-\frac{1}{2}+\frac{1}{2}\sqrt{5}}{\sqrt{-2+2\sqrt{5}}}\right), \; e^{\left(2\,\frac{-1/2+1/2\sqrt{5}}{\sqrt{-2+2\sqrt{5}}\sqrt{1+4\frac{(-1/2+1/2\sqrt{5})^2}{-2+2\sqrt{5}}}}\right)} \right] \right\}$$

```
> evalf( % );
```

$$\{[2.475353222, \; 1.855276958],$$

$$[.6662394325, \; 1.855276958]\}$$

Since $f$ is periodic, it has, of course, infinitely many inflection points. You can obtain these by shifting the two inflection points above horizontally by multiples of $2\pi$.

## A Taylor Approximation

This section illustrates how you can use Maple to analyse the error term in a Taylor approximation. Below is Taylor's formula.

```
> taylor( f(x), x=a );
```

$$f(a) + D(f)(a)\,(x - a) + \frac{1}{2}\,(D^{(2)})(f)(a)\,(x - a)^2 + \frac{1}{6}\,(D^{(3)})(f)(a)$$

$$(x - a)^3 + \frac{1}{24}\,(D^{(4)})(f)(a)\,(x - a)^4 + \frac{1}{120}\,(D^{(5)})(f)(a)\,(x - a)^5 +$$

$$O((x - a)^6)$$

You can use it to find a polynomial approximation of a function $f$ near $x = a$.

```
> f := x -> exp( sin(x) );
```

$$f := x \rightarrow e^{\sin(x)}$$

```
> a := Pi;
```

$$a := \pi$$

```
> taylor( f(x), x=a );
```

$$1 - (x - \pi) + \frac{1}{2}\,(x - \pi)^2 - \frac{1}{8}\,(x - \pi)^4 + \frac{1}{15}\,(x - \pi)^5 +$$

$$O((x - \pi)^6)$$

Before you can plot the Taylor approximation, you must convert it from a series to a polynomial.

```
> poly := convert( %, polynom );
```

$$poly := 1 - x + \pi + \frac{1}{2}\,(x - \pi)^2 - \frac{1}{8}\,(x - \pi)^4 + \frac{1}{15}\,(x - \pi)^5$$

Now plot the function $f$ along with poly.

```
> plot( [f(x), poly], x=0..10, view=[0..10, 0..3] );
```

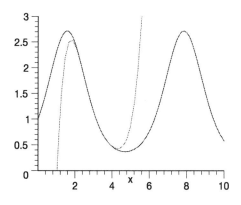

The expression $(1/6!) f^{(6)}(\xi)(x-a)^6$ gives the error of the approximation, where $\xi$ is some number between $x$ and $a$. The sixth derivative of $f$ is

```
> diff( f(x), x$6 );
```

$$-\sin(x)\,e^{\sin(x)} + 16\,\cos(x)^2\,e^{\sin(x)} - 15\,\sin(x)^2\,e^{\sin(x)}$$
$$+ 75\,\sin(x)\cos(x)^2\,e^{\sin(x)} - 20\,\cos(x)^4\,e^{\sin(x)}$$
$$- 15\,\sin(x)^3\,e^{\sin(x)} + 45\,\sin(x)^2\cos(x)^2\,e^{\sin(x)}$$
$$- 15\,\sin(x)\cos(x)^4\,e^{\sin(x)} + \cos(x)^6\,e^{\sin(x)}$$

Define the function $f6$ to be that derivative.

```
> f6 := unapply( %, x );
```

$$f6 := x \rightarrow -\sin(x)\,e^{\sin(x)} + 16\cos(x)^2\,e^{\sin(x)}$$
$$- 15\sin(x)^2\,e^{\sin(x)} + 75\sin(x)\cos(x)^2\,e^{\sin(x)}$$
$$- 20\cos(x)^4\,e^{\sin(x)} - 15\sin(x)^3\,e^{\sin(x)}$$
$$+ 45\sin(x)^2\cos(x)^2\,e^{\sin(x)} - 15\sin(x)\cos(x)^4\,e^{\sin(x)}$$
$$+ \cos(x)^6\,e^{\sin(x)}$$

The following is the error in the approximation.

```
> err := 1/6! * f6(xi) * (x - a)^6;
```

$$err := \frac{1}{720}(-\sin(\xi)\,e^{\sin(\xi)} + 16\cos(\xi)^2\,e^{\sin(\xi)}$$
$$- 15\sin(\xi)^2\,e^{\sin(\xi)} + 75\sin(\xi)\cos(\xi)^2\,e^{\sin\,(\xi)}$$
$$- 20\cos(\xi)^4\,e^{\sin(\xi)} - 15\sin(\xi)^3\,e^{\sin(\xi)}$$
$$+ 45\sin(\xi)^2\cos(\xi)^2\,e^{\sin(\xi)} - 15\sin(\xi)\cos(\xi)^4\,e^{\sin(\xi)}$$
$$+ \cos(\xi)^6\,e^{\sin(\xi)})(x - \pi)^6$$

The plot above indicates that the error is small (in absolute value) for $x$ between 2 and 4.

```
> plot3d( abs( err ), x=2..4, xi=2..4,
>    style=patch, axes=boxed );
```

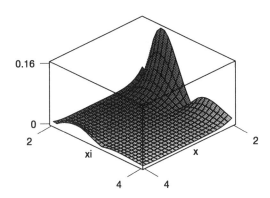

To find the size of the error, you need a full analysis of the expression err for $x$ between 2 and 4 and $\xi$ between $a$ and $x$; that is, on the two closed regions bounded by $x = 2$, $x = 4$, $\xi = a$, and $\xi = x$. The curve command from the plottools package can illustrate these two regions.

```
> with(plots): with(plottools):
> display( curve( [ [2,2], [2,a], [4,a], [4,4], [2,2] ] ),
>          labels=[x, xi] );
```

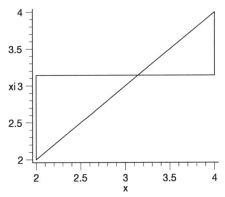

The partial derivatives of err help you find extrema of err inside the two regions. Then you need to attend to the four boundaries. The two partial derivatives of err are

```
> err_x := diff(err, x);
```

$$err\_x := \frac{1}{120}(-\sin(\xi)\, e^{\sin(\xi)} + 16\cos(\xi)^2\, e^{\sin(\xi)})$$

$$-15 \sin(\xi)^2 \, e^{\sin(\xi)} + 75 \sin(\xi) \cos(\xi)^2 \, e^{\sin(\xi)}$$
$$-20 \cos(\xi)^4 \, e^{\sin(\xi)} - 15 \sin(\xi)^3 \, e^{\sin(\xi)}$$
$$+45 \sin(\xi)^2 \cos(\xi)^2 \, e^{\sin(\xi)} - 15 \sin(\xi) \cos(\xi)^4 \, e^{\sin(\xi)}$$
$$+\cos(\xi)^6 \, e^{\sin(\xi)})(x - \pi)^5$$

```
> err_xi := diff(err, xi);
```

$$err\_xi := \frac{1}{720}(-\cos(\xi) \, e^{\sin(\xi)} - 63 \sin(\xi) \cos(\xi) \, e^{\sin(\xi)}$$
$$+91 \cos(\xi)^3 \, e^{\sin(\xi)} - 210 \sin(\xi)^2 \cos(\xi) \, e^{\sin(\xi)}$$
$$+245 \sin(\xi) \cos(\xi)^3 \, e^{\sin(\xi)} - 35 \cos(\xi)^5 \, e^{\sin(\xi)}$$
$$-105 \sin(\xi)^3 \cos(\xi) \, e^{\sin(\xi)} + 105 \sin(\xi)^2 \cos(\xi)^3 \, e^{\sin(\xi)}$$
$$-21 \sin(\xi) \cos(\xi)^5 \, e^{\sin(\xi)} + \cos(\xi)^7 \, e^{\sin(\xi)})(x - \pi)^6$$

The two partial derivatives are zero at a critical point.

```
> sol := solve( {err_x=0, err_xi=0}, {x, xi} );
```

$$sol := \{\xi = \xi, \, x = \pi\}$$

The error is zero at this critical point.

```
> eval( err, sol );
```

$$0$$

You need to collect a set of critical values. The largest critical value then bounds the maximal error.

```
> critical := { % };
```

$$critical := \{0\}$$

The partial derivative err_xi is zero at a critical point on either of the two boundaries at $x = 2$ and $x = 4$.

```
> sol := { solve( err_xi=0, xi ) };
```

$$sol := \{\frac{1}{2}\pi, \, \arctan(\text{RootOf}(-56 - 161 \, \_Z + 129 \, \_Z^2$$
$$+308 \, \_Z^3 + 137 \, \_Z^4 + 21 \, \_Z^5 + \_Z^6), \text{RootOf}( \, \_Z^2 +$$
$$\text{RootOf}(-56 - 161 \, \_Z + 129 \, \_Z^2 + 308 \, \_Z^3 + 137 \, \_Z^4$$
$$+21 \, \_Z^5 + \_Z^6)^2 - 1))\}$$

Only the real solutions are interesting.

```
> select( type, sol, realcons );
```

$$\left\{ \frac{1}{2} \pi \right\}$$

You should check the solution set by plotting the function.

```
> plot( eval(err_xi, x=2), xi=2..4 );
```

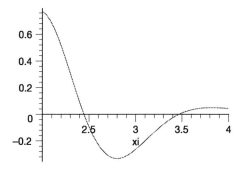

Two solutions to `err_xi=0` seem to exist between 2 and 4 where `solve` found none: $\pi/2$ is less than 2. Thus, you must use numerical methods. If $x = 2$, then $\xi$ should be in the interval from 2 to $a$.

```
> sol := fsolve( eval(err_xi, x=2), xi, 2..a );
```

$$sol := 2.446729125$$

At that point the error is

```
> eval( err, {x=2, xi=sol});
```

$$.07333000221\,(2-\pi)^6$$

Now add this value to the set of critical values.

```
> critical := critical union {%};
```

$$critical := \{.07333000221\,(2-\pi)^6,\ 0\}$$

If $x = 4$ then $\xi$ should be between $a$ and 4.

```
> sol := fsolve( eval(err_xi, x=4), xi, a..4 );
```

$$sol := 3.467295314$$

At that point, the error is

```
> eval( err, {x=4, xi=sol} );
```

$$-.01542298121\,(4-\pi)^6$$

```
> critical := critical union {%};
```

$$critical :=$$
$$\{-.01542298121\,(4-\pi)^6,\ .07333000221\,(2-\pi)^6,\ 0\}$$

At the $\xi = a$ boundary, the error is

```
> B := eval( err, xi=a );
```

$$B := -\frac{1}{240}\,(x-\pi)^6$$

The derivative, $B1$, of $B$ is zero at a critical point.

```
> B1 := diff( B, x );
```

$$B1 := -\frac{1}{40}\,(x-\pi)^5$$

```
> sol := { solve( B1=0, x ) };
```

$$sol := \{\pi\}$$

At the critical point the error is

```
> eval( B, x=sol[1] );
```

$$0$$

```
> critical := critical union { % };
```

$$critical :=$$
$$\{-.01542298121\,(4-\pi)^6,\ .07333000221\,(2-\pi)^6,\ 0\}$$

At the last boundary, $\xi = x$, the error is

```
> B := eval( err, xi=x );
```

$$B := \frac{1}{720}(-\sin(x)\,e^{\sin(x)} + 16\cos(x)^2\,e^{\sin(x)} - 15\sin(x)^2\,e^{\sin(x)}$$

$$+\,75\sin(x)\cos(x)^2\,e^{\sin(x)} - 20\cos(x)^4\,e^{\sin(x)}$$

$$-\,15\sin(x)^3\,e^{\sin(x)} + 45\sin(x)^2\cos(x)^2\,e^{\sin(x)}$$

$$-\,15\sin(x)\cos(x)^4\,e^{\sin(x)} + \cos(x)^6\,e^{\sin(x)})(x-\pi)^6$$

Again, you need to find where the derivative is zero.

```
> B1 := diff( B, x );
```

$$B1 := \frac{1}{720}(-\cos(x)\,e^{\sin(x)} - 63\sin(x)\cos(x)\,e^{\sin(x)}$$

$$+\,91\cos(x)^3\,e^{\sin(x)} - 210\sin(x)^2\cos(x)\,e^{\sin(x)}$$

$$+\,245\sin(x)\cos(x)^3\,e^{\sin(x)} - 35\cos(x)^5\,e^{\sin(x)}$$

$$-\,105\sin(x)^3\cos(x)\,e^{\sin(x)} + 105\sin(x)^2\cos(x)^3\,e^{\sin(x)}$$

$$-\,21\sin(x)\cos(x)^5\,e^{\sin(x)} + \cos(x)^7\,e^{\sin(x)})(x-\pi)^6 + \frac{1}{120}($$

$$-\sin(x)\,e^{\sin(x)} + 16\cos(x)^2\,e^{\sin(x)} - 15\sin(x)^2\,e^{\sin(x)}$$

$$+\,75\sin(x)\cos(x)^2\,e^{\sin(x)} - 20\cos(x)^4\,e^{\sin(x)}$$

$$-\,15\sin(x)^3\,e^{\sin(x)} + 45\sin(x)^2\cos(x)^2\,e^{\sin(x)}$$

$$-\,15\sin(x)\cos(x)^4\,e^{\sin(x)} + \cos(x)^6\,e^{\sin(x)})(x-\pi)^5$$

```
> sol := { solve( B1=0, x ) };
```

$$sol := \{\pi\}$$

Checking the solution by plotting is a good idea.

```
> plot( B1, x=2..4 );
```

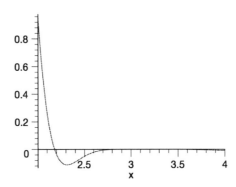

The plot of $B1$ indicates that a solution between 2.1 and 2.3 exists. solve cannot find that solution, so you must resort to numerical methods again.

```
> fsolve( B1=0, x, 2.1..2.3 );
```

$$2.180293062$$

Add the numerical solution to the set of symbolic solutions.

```
> sol := sol union { % };
```

$$sol := \{\pi, 2.180293062\}$$

The following is the set of extremal errors at the $\xi = x$ boundary.

```
> { seq( B, x=sol ) };
```

$$\{0, .04005698601\,(2.180293062 - \pi)^6\}$$

Now enlarge the set of large errors.

```
> critical := critical union %;
```

$$critical := \{-.01542298121\,(4 - \pi)^6,$$
$$.07333000221\,(2 - \pi)^6, 0,$$
$$.04005698601\,(2.180293062 - \pi)^6\}$$

Finally, you must add the error at the four corners to the set of critical values.

```
> critical := critical union
>    { eval( err, {xi=2, x=2} ),
>        eval( err, {xi=2, x=4} ),
```

```
>       eval( err, {xi=4, x=2} ),
>       eval( err, {xi=4, x=4} ) };
```

$critical := \{-.01542298121\,(4 - \pi)^6,$

$.07333000221\,(2 - \pi)^6,\ 0,\ \dfrac{1}{720}(-\sin(2)\,e^{\sin(2)}$

$+\,16\cos(2)^2\,e^{\sin(2)} - 15\sin(2)^2\,e^{\sin(2)}$

$+\,75\sin(2)\cos(2)^2\,e^{\sin(2)} - 20\cos(2)^4\,e^{\sin(2)}$

$-\,15\sin(2)^3\,e^{\sin(2)} + 45\sin(2)^2\cos(2)^2\,e^{\sin(2)}$

$-\,15\sin(2)\cos(2)^4\,e^{\sin(2)} + \cos(2)^6\,e^{\sin(2)})(4 - \pi)^6,\ \dfrac{1}{720}($

$-\,\sin(2)\,e^{\sin(2)} + 16\cos(2)^2\,e^{\sin(2)} - 15\sin(2)^2\,e^{\ \sin(2)}$

$+\,75\sin(2)\cos(2)^2\,e^{\sin(2)} - 20\cos(2)^4\,e^{\sin(2)}$

$-\,15\sin(2)^3\,e^{\sin(2)} + 45\sin(2)^2\cos(2)^2\,e^{\sin(2)}$

$-\,15\sin(2)\cos(2)^4\,e^{\sin(2)} + \cos(2)^6\,e^{\sin(2)})(2 - \pi)^6,$

$.04005698601\,(2.180293062 - \pi)^6,\ \dfrac{1}{720}(-\sin(4)\,e^{\sin(4)}$

$+\,16\cos(4)^2\,e^{\sin(4)} - 15\sin(4)^2\,e^{\sin(4)}$

$+\,75\sin(4)\cos(4)^2\,e^{\sin(4)} - 20\cos(4)^4\,e^{\sin(4)}$

$-\,15\sin(4)^3\,e^{\sin(4)} + 45\sin(4)^2\cos(4)^2\,e^{\sin(4)}$

$-\,15\sin(4)\cos(4)^4\,e^{\sin(4)} + \cos(4)^6\,e^{\sin(4)})(4 - \pi)^6,\ \dfrac{1}{720}($

$-\,\sin(4)\,e^{\sin(4)} + 16\cos(4)^2\,e^{\sin(4)} - 15\sin(4)^2\,e^{\ \sin(4)}$

$+\,75\sin(4)\cos(4)^2\,e^{\sin(4)} - 20\cos(4)^4\,e^{\sin(4)}$

$-\,15\sin(4)^3\,e^{\sin(4)} + 45\sin(4)^2\cos(4)^2\,e^{\sin(4)}$

$-\,15\sin(4)\cos(4)^4\,e^{\sin(4)} + \cos(4)^6\,e^{\sin(4)})(2 - \pi)^6\}$

Now all you need to do is find the maximum of the absolute values of the elements of `critical`. First, map the abs command onto the elements of `critical`.

```
> map( abs, critical );
```

$$\{.01542298121 \, (4 - \pi)^6, \; -\frac{1}{720}(-\sin(4) \, e^{\sin(4)}$$

$$+ 16 \cos(4)^2 \, e^{\sin(4)} - 15 \sin(4)^2 \, e^{\sin(4)}$$

$$+ 75 \sin(4) \cos(4)^2 \, e^{\sin(4)} - 20 \cos(4)^4 \, e^{\sin(4)}$$

$$- 15 \sin(4)^3 \, e^{\sin(4)} + 45 \sin(4)^2 \cos(4)^2 \, e^{\sin(4)}$$

$$- 15 \sin(4) \cos(4)^4 \, e^{\sin(4)} + \cos(4)^6 \, e^{\sin(4)})(4 - \pi)^6, \; -\frac{1}{720}($$

$$- \sin(2) \, e^{\sin(2)} + 16 \cos(2)^2 \, e^{\sin(2)} - 15 \sin(2)^2 \, e^{\sin(2)}$$

$$+ 75 \sin(2) \cos(2)^2 \, e^{\sin(2)} - 20 \cos(2)^4 \, e^{\sin(2)}$$

$$- 15 \sin(2)^3 \, e^{\sin(2)} + 45 \sin(2)^2 \cos(2)^2 \, e^{\sin(2)}$$

$$- 15 \sin(2) \cos(2)^4 \, e^{\sin(2)} + \cos(2)^6 \, e^{\sin(2)})(2 - \pi)^6, \; -\frac{1}{720}($$

$$- \sin(2) \, e^{\sin(2)} + 16 \cos(2)^2 \, e^{\sin(2)} - 15 \sin(2)^2 \, e^{\sin(2)}$$

$$+ 75 \sin(2) \cos(2)^2 \, e^{\sin(2)} - 20 \cos(2)^4 \, e^{\sin(2)}$$

$$- 15 \sin(2)^3 \, e^{\sin(2)} + 45 \sin(2)^2 \cos(2)^2 \, e^{\sin(2)}$$

$$- 15 \sin(2) \cos(2)^4 \, e^{\sin(2)} + \cos(2)^6 \, e^{\sin(2)})(4 - \pi)^6, \; -\frac{1}{720}($$

$$- \sin(4) \, e^{\sin(4)} + 16 \cos(4)^2 \, e^{\sin(4)} - 15 \sin(4)^2 \, e^{\sin(4)}$$

$$+ 75 \sin(4) \cos(4)^2 \, e^{\sin(4)} - 20 \cos(4)^4 \, e^{\sin(4)}$$

$$- 15 \sin(4)^3 \, e^{\sin(4)} + 45 \sin(4)^2 \cos(4)^2 \, e^{\sin(4)}$$

$$- 15 \sin(4) \cos(4)^4 \, e^{\sin(4)} + \cos(4)^6 \, e^{\sin(4)})(2 - \pi)^6,$$

$$.07333000221 \, (2 - \pi)^6, \; 0,$$

$$.04005698601 \, (2.180293062 - \pi)^6\}$$

Then find the maximal element. The `max` command expects a sequence of numbers, so you must use the `op` command to turn the set of values into a sequence.

```
> max_error := max( op(%) );
```

$$max\_error := .07333000221 \, (2 - \pi)^6$$

Approximately, this number is

```
> evalf( max_error );
```

$$.1623112756$$

You can now plot $f$, its Taylor approximation, and a pair of curves indicating the error band.

```
> plot( [ f(x), poly, f(x)+max_error, f(x)-max_error ],
>         x=2..4,
>         color=[ red, blue, brown, brown ] );
```

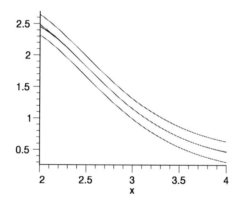

The plot shows that the actual error stays well inside the error estimate.

## The Integral

The integral of a function can be considered as a measure of the area between the $x$-axis and the graph of the function. The definition of the Riemann integral relies on this graphical interpretation of the integral.

```
> f := x ->   1/2 + sin(x);
```

$$f := x \rightarrow \frac{1}{2} + \sin(x)$$

Here, the leftbox command from the student package draws the graph of $f$ along with 6 boxes. The height of each box is the value of $f$ evaluated at the left-hand side of the box.

```
> with(student):
```

```
> leftbox( f(x), x=0..10, 6 );
```

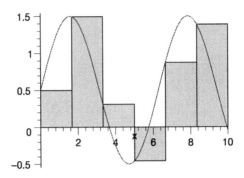

The `leftsum` command calculates the area of the boxes.

```
> leftsum( f(x), x=0..10, 6 );
```

$$\frac{5}{3} \left( \sum_{i=0}^{5} \left( \frac{1}{2} + \sin\left(\frac{5}{3}i\right) \right) \right)$$

Approximately, this number is

```
> evalf( % );
```

$$6.845601766$$

The approximation of the area gets better as you use more and more boxes.

```
> boxes := [ seq( i^2, i=3..14 ) ];
```

$$boxes :=$$

$$[9, 16, 25, 36, 49, 64, 81, 100, 121, 144, 169, 196]$$

For each number in the list boxes, calculate the value of `leftsum`.

```
> seq( evalf( leftsum( f(x), x=0..10, n ) ), n=boxes );
```

$$6.948089404, \ 6.948819106, \ 6.923289160, \ 6.902789476,$$

$$6.888196449, \ 6.877830055, \ 6.870316621,$$

$$6.864739770, \ 6.860504862, \ 6.857222009,$$

$$6.854630207, \ 6.852550663$$

You can specify a title to the `leftbox` plot. For example, you can use the value of the `leftsum` as the title.

```
> S := seq( leftbox( f(x), x=0..10, n,
>     title=convert( evalf( leftsum( f(x), x=0..10, n ) ),
>                        string ) ),
>        n=boxes ):
```

The display command from the plots package can show the sequence S of plots as an animation.

```
> with(plots):
```

```
> display( S, insequence=true );
```

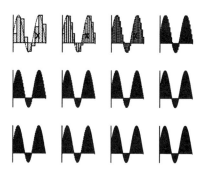

In the limit, as the number of boxes gets large, you obtain the definite integral.

```
> Int( f(x), x=0..10 );
```

$$\int_0^{10} \frac{1}{2} + \sin(x) \, dx$$

The value of the integral is

```
> value( % );
```

$$6 - \cos(10)$$

and in floating-point numbers, this value is approximately

```
> evalf( % );
```

$$6.839071529$$

The indefinite integral of $f$ is

```
> Int( f(x), x );
```

$$\int \frac{1}{2} + \sin(x) \, dx$$

```
> value( % );
```

$$\frac{1}{2}x - \cos(x)$$

Define the function $F$ to be the anti-derivative.

```
> F := unapply( %, x );
```

$$F := x \rightarrow \frac{1}{2}x - \cos(x)$$

Choose the constant of integration so that $F(0) = 0$.

```
> F(x) - F(0);
```

$$\frac{1}{2}x - \cos(x) + 1$$

```
> F := unapply( %, x );
```

$$F := x \rightarrow \frac{1}{2}x - \cos(x) + 1$$

If you plot $F$ and the left-boxes together, you can see that $F$ increases more when the corresponding box is larger.

```
> display( [ plot( F(x), x=0..10, color=blue ),
>                leftbox( f(x), x=0..10, 14 ) ] );
```

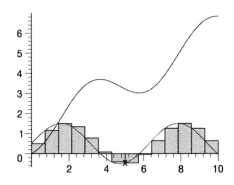

The student package also contains commands for drawing and summing boxes evaluated at the right-hand side or at the midpoint of the box.

## Mixed Partial Derivatives

This section describes the D operator for derivatives and gives an example of a function whose mixed partial derivatives are different.
Consider the following function.

```
> f := (x,y) -> x * y * (x^2-y^2) / (x^2+y^2);
```

$$f := (x, y) \rightarrow \frac{x\,y\,(x^2 - y^2)}{x^2 + y^2}$$

The function $f$ is not defined at $(0, 0)$.

```
> f(0,0);
```

```
Error, (in f) division by zero
```

At $(x, y) = (r\cos(\theta), r\sin(\theta))$ the function value is

```
> f( r*cos(theta), r*sin(theta) );
```

$$\frac{r^2 \cos(\theta) \sin(\theta)\,(r^2 \cos(\theta)^2 - r^2 \sin(\theta)^2)}{r^2 \cos(\theta)^2 + r^2 \sin(\theta)^2}$$

As $r$ tends to zero so does the function value.

```
> Limit( %, r=0 );
```

$$\lim_{r \to 0} \frac{r^2 \cos(\theta) \sin(\theta)\,(r^2 \cos(\theta)^2 - r^2 \sin(\theta)^2)}{r^2 \cos(\theta)^2 + r^2 \sin(\theta)^2}$$

```
> value( % );
```

$$0$$

Thus, you may extend $f$ as a continuous function by defining it to be zero at $(0, 0)$.

```
> f(0,0) := 0;
```

$$f(0, 0) := 0$$

The above assignment places an entry in $f$'s remember table. Here is the graph of $f$.

```
> plot3d( f, -3..3, -3..3 );
```

The partial derivative of $f$ with respect to its first parameter, $x$, is

```
> fx := D[1](f);
```

$$fx := (x, y) \to \frac{y\,(x^2 - y^2)}{x^2 + y^2} + 2\,\frac{x^2\,y}{x^2 + y^2} - 2\,\frac{x^2\,y\,(x^2 - y^2)}{(x^2 + y^2)^2}$$

This formula does not hold at $(0, 0)$.

```
> fx(0,0);
```

```
Error, (in fx) division by zero
```

Therefore, you must use the limit definition of the derivative.

```
> fx(0,0) := limit( ( f(h,0) - f(0,0) )/h, h=0 );
```

$$fx(0, 0) := 0$$

At $(x, y) = (r \cos(\theta), r \sin(\theta))$ the value of $fx$ is

```
> fx( r*cos(theta), r*sin(theta) );
```

$$\frac{r \sin(\theta)\,(r^2 \cos(\theta)^2 - r^2 \sin(\theta)^2)}{r^2 \cos(\theta)^2 + r^2 \sin(\theta)^2} + 2\,\frac{r^3 \cos(\theta)^2 \sin(\theta)}{r^2 \cos(\theta)^2 + r^2 \sin(\theta)^2}$$
$$- 2\,\frac{r^3 \cos(\theta)^2 \sin(\theta)\,(r^2 \cos(\theta)^2 - r^2 \sin(\theta)^2)}{(r^2 \cos(\theta)^2 + r^2 \sin(\theta)^2)^2}$$

```
> combine( % );
```

$$\frac{3}{4}\,r \sin(3\,\theta) - \frac{1}{4}\,r \sin(5\,\theta)$$

As the distance $r$ from $(x, y)$ to $(0, 0)$ tends to zero, so does $|fx(x, y) - fx(0, 0)|$.

```
> Limit( abs( % - fx(0,0) ), r=0 );
```

$$\lim_{r \to 0} \left| -\frac{3}{4}\,r \sin(3\,\theta) + \frac{1}{4}\,r \sin(5\,\theta) \right|$$

```
> value( % );
```

$$0$$

Hence, $fx$ is continuous at $(0, 0)$.

By symmetry, the same arguments apply to the derivative of $f$ with respect to its second parameter, $y$.

```
> fy := D[2](f);
```

$$fy := (x, y) \to \frac{x\,(x^2 - y^2)}{x^2 + y^2} - 2\,\frac{x\,y^2}{x^2 + y^2} - 2\,\frac{x\,y^2\,(x^2 - y^2)}{(x^2 + y^2)^2}$$

```
> fy(0,0) := limit( ( f(0,k) - f(0,0) )/k, k=0 );
```

$$fy(0, 0) := 0$$

Here is a mixed second derivative of $f$.

```
> fxy := D[1,2](f);
```

$$fxy := (x, y) \rightarrow \frac{x^2 - y^2}{x^2 + y^2} + 2\frac{x^2}{x^2 + y^2} - 2\frac{x^2(x^2 - y^2)}{(x^2 + y^2)^2}$$

$$- 2\frac{y^2}{x^2 + y^2} - 2\frac{y^2(x^2 - y^2)}{(x^2 + y^2)^2} + 8\frac{x^2 y^2(x^2 - y^2)}{(x^2 + y^2)^3}$$

Again, the formula does not hold at $(0, 0)$.

```
> fxy(0,0);
```

Error, (in fxy) division by zero

The limit definition is

```
> Limit( ( fx(0,k) - fx(0,0) )/k, k=0 );
```

$$\lim_{k \to 0} -1$$

```
> fxy(0,0) := value( % );
```

$$fxy(0, 0) := -1$$

The other mixed second derivative is

```
> fyx := D[2,1](f);
```

$$fyx := (x, y) \rightarrow \frac{x^2 - y^2}{x^2 + y^2} + 2\frac{x^2}{x^2 + y^2} - 2\frac{x^2(x^2 - y^2)}{(x^2 + y^2)^2}$$

$$- 2\frac{y^2}{x^2 + y^2} - 2\frac{y^2(x^2 - y^2)}{(x^2 + y^2)^2} + 8\frac{x^2 y^2(x^2 - y^2)}{(x^2 + y^2)^3}$$

At $(0, 0)$, you need to use the limit definition.

```
> Limit( ( fy(h, 0) - fy(0,0) )/h, h=0 );
```

$$\lim_{h \to 0} 1$$

```
> fyx(0,0) := value( % );
```

$$fyx(0, 0) := 1$$

Note that the two mixed partial derivatives are different at $(0, 0)$.

```
> fxy(0,0) <> fyx(0,0);
```

$$-1 \neq 1$$

```
> evalb( % );
```

*true*

The mixed partial derivatives are equal only if they are continuous. If you plot fxy, you can see that it is not continuous at $(0, 0)$.

```
> plot3d( fxy, -3..3, -3..3 );
```

Maple can help you with many other problems from introductory calculus. The ?student help page is a good source of inspiration.

## 6.2 Ordinary Differential Equations

Maple provides you with a varied set of tools for solving, manipulating, and plotting ordinary differential equations and systems of differential equations.

### The dsolve Command

The most commonly used command for investigating the behavior of ordinary differential equations (ODEs) within Maple is dsolve. You can use this general-purpose tool to obtain both closed form and numerical solutions to a wide variety of ODEs. This is the basic syntax of dsolve.

```
dsolve(eqns, vars)
```

Here *eqns* is a set of differential equations and initial values, and *vars* is a set of variables that need solving.

Here is a differential equation and an initial condition.

```
> eq := diff(v(t),t)+2*t = 0;
```

$$eq := \left( \frac{\partial}{\partial t} v(t) \right) + 2t = 0$$

```
> ini := v(1) = 5;
```

$$ini := v(1) = 5$$

Use dsolve to obtain the solution.

```
> dsolve( {eq, ini}, {v(t)} );
```

$$v(t) = -t^2 + 6$$

If you omit some or all of the initial conditions, then dsolve returns a solution containing arbitrary constants of the form _Cnumber.

```
> eq := diff(y(x),x$2) - y(x) = 1;
```

$$eq := \left( \frac{\partial^2}{\partial x^2} y(x) \right) - y(x) = 1$$

```
> dsolve( {eq}, {y(x)} );
```

$$y(x) = -1 + \_C1 \sinh(x) + \_C2 \cosh(x)$$

To specify initial conditions for the derivative of a function, use the following notation.

```
D(fcn)(var_value) = value
(D@@n)(fcn)(var_value) = value
```

Here is a differential equation and some initial conditions involving the derivative.

```
> de1 := diff(y(t),t$2) + 5*diff(y(t),t) + 6*y(t) = 0;
```

$$de1 := \left( \frac{\partial^2}{\partial t^2} y(t) \right) + 5 \left( \frac{\partial}{\partial t} y(t) \right) + 6 y(t) = 0$$

```
> ini := y(0)=0, D(y)(0)=1;
```

$$ini := y(0) = 0, \ D(y)(0) = 1$$

Again, use dsolve to find the solution.

```
> dsolve( {de1, ini}, {y(t)} );
```

$$y(t) = -e^{(-3t)} + e^{(-2t)}$$

Additionally, dsolve may return a solution in parametric form, [x=f(_T), y(x)=g(_T)], where _T is the parameter.

**The explicit Option**   Maple may return the solution to a differential equation in implicit form.

```
> de2 := diff(y(x),x$2) = (ln(y(x))+1)*diff(y(x),x);
```

$$de2 := \frac{\partial^2}{\partial x^2} y(x) = (\ln(y(x)) + 1) \left( \frac{\partial}{\partial x} y(x) \right)$$

```
> dsolve( {de2}, {y(x)} );
```

$$\int^{y(x)} \frac{1}{\_a \ln(\_a) + \_C1} \, d\_a - x - \_C2 = 0, \; y(x) = \_C2$$

Use the explicit option to tell Maple to look for an explicit solution for the first result.

```
> dsolve( {de2}, {y(x)}, explicit );
```

$$y(x) = \text{RootOf} \left( \int^{\_Z} \frac{1}{\_c \ln(\_c) + \_C1} \, d\_c - x - \_C2 \right),$$

$$y(x) = \_C2$$

However, in some cases, Maple may not be able to find an explicit solution. There is also an implicit option to force answers in implicit form.

**The method=laplace Option**   Applying Laplace transform methods to differential equations often reduces the complexity of the problem. The transform maps the differential equations into algebraic equations, which are much easier to solve. The difficulty is in the transformation of the equations to the new domain, and especially the transformation of the solutions back.

The Laplace transform method can handle linear ODEs of arbitrary order, and some cases of linear ODEs with non-constant coefficients, provided that Maple can find the transforms. This method can also deal with systems of coupled equations.

Consider the following problem in classical dynamics. Two weights with masses $m$ and $\alpha m$, respectively, rest on a frictionless plane joined by a spring with spring constant $k$. What are the trajectories of each weight if the first weight is subject to a unit step force $u(t)$ at time $t = 1$? First, set up the differential equations that govern the system. Newton's Second Law governs the motion of the first weight, and hence, the mass $m$ times the acceleration must equal the sum of the forces that you apply to the first weight, including the external force $u(t)$.

```
> eqn1 :=
>     alpha*m*diff(x[1](t),t$2) = k*(x[2](t) - x[1](t)) + u(t);
```

$$eqn1 := \alpha\, m \left( \frac{\partial^2}{\partial t^2} x_1(t) \right) = k\,(x_2(t) - x_1(t)) + u(t)$$

Similarly for the second weight.

```
> eqn2 := m*diff(x[2](t),t$2) = k*(x[1](t) - x[2](t));
```

$$eqn2 := m \left( \frac{\partial^2}{\partial t^2} x_2(t) \right) = k\,(x_1(t) - x_2(t))$$

Apply a unit step force to the first weight at $t = 1$.

```
> u := t -> Heaviside(t-1);
```

$$u := t \to \mathrm{Heaviside}(t - 1)$$

At time $t = 0$, both masses are at rest at their respective locations.

```
> ini := x[1](0) = 2, D(x[1])(0) = 0,
>          x[2](0) = 0, D(x[2])(0) = 0 ;
```

$$ini := x_1(0) = 2,\ D(x_1)(0) = 0,\ x_2(0) = 0,\ D(x_2)(0) = 0$$

Solve the problem using Laplace transform methods.

```
> dsolve( {eqn1, eqn2, ini}, {x[1](t), x[2](t)},
>     method=laplace );
```

$$\left\{ x_2(t) = 2\,\alpha\, k \left( \frac{1}{k\,(\alpha + 1)} - \frac{\cos\left( \sqrt{\frac{k\,(\alpha + 1)}{\alpha\, m}}\, t \right)}{k\,(\alpha + 1)} \right) + k \left( \right. \right.$$

$$-\frac{\mathrm{Heaviside}(t - 1)\,\alpha\, m}{k^2\,(\alpha + 1)^2} + \frac{1}{2}\frac{\mathrm{Heaviside}(t - 1)\,t^2}{k\,(\alpha + 1)}$$

$$-\frac{\mathrm{Heaviside}(t - 1)\,t}{k\,(\alpha + 1)} + \frac{1}{2}\frac{\mathrm{Heaviside}(t - 1)}{k\,(\alpha + 1)}$$

$$\left. \left. + \frac{\mathrm{Heaviside}(t - 1)\,\alpha\, m \cos\left( \sqrt{\frac{k\,(\alpha + 1)}{\alpha\, m}}\,(t - 1) \right)}{k^2\,(\alpha + 1)^2} \right) \middle/ m, \right.$$

$$x_1(t) = 2\cos\left(\sqrt{\frac{k(\alpha+1)}{\alpha m}}\,t\right)$$

$$+\,2\alpha k\left(\frac{1}{k(\alpha+1)} - \frac{\cos\left(\sqrt{\frac{k(\alpha+1)}{\alpha m}}\,t\right)}{k(\alpha+1)}\right)$$

$$+\,\frac{\text{Heaviside}(t-1)}{k(\alpha+1)}$$

$$-\,\frac{\text{Heaviside}(t-1)\cos\left(\sqrt{\frac{k(\alpha+1)}{\alpha m}}(t-1)\right)}{k(\alpha+1)} + k\left(\vphantom{\frac{1}{1}}\right.$$

$$-\,\frac{\text{Heaviside}(t-1)\,\alpha m}{k^2(\alpha+1)^2} + \frac{1}{2}\frac{\text{Heaviside}(t-1)\,t^2}{k(\alpha+1)}$$

$$-\,\frac{\text{Heaviside}(t-1)\,t}{k(\alpha+1)} + \frac{1}{2}\frac{\text{Heaviside}(t-1)}{k(\alpha+1)}$$

$$+\,\frac{\text{Heaviside}(t-1)\,\alpha m\cos\left(\sqrt{\frac{k(\alpha+1)}{\alpha m}}(t-1)\right)}{k^2(\alpha+1)^2}\left.\vphantom{\frac{1}{1}}\right)\Big/m\Bigg\}$$

Evaluate the result at values for the constants.

```
> ans := eval( %, {alpha=1/10, m=1, k=1} );
```

$$ans := \{x_1(t) = \frac{20}{11}\cos(\sqrt{11}\,t) + \frac{2}{11} + \frac{155}{121}\text{Heaviside}(t-1)$$

$$-\,\frac{100}{121}\text{Heaviside}(t-1)\cos(\sqrt{11}\,(t-1))$$

$$+\,\frac{5}{11}\text{Heaviside}(t-1)\,t^2 - \frac{10}{11}\text{Heaviside}(t-1)\,t, x_2(t) =$$

$$\frac{2}{11} - \frac{2}{11}\cos(\sqrt{11}\,t) + \frac{45}{121}\text{Heaviside}(t-1)$$

$$+\,\frac{5}{11}\text{Heaviside}(t-1)\,t^2 - \frac{10}{11}\text{Heaviside}(t-1)\,t$$

$$+\,\frac{10}{121}\text{Heaviside}(t-1)\cos(\sqrt{11}\,(t-1))\}$$

You can turn the above solution into two functions, say $y_1(t)$ and $y_2(t)$, as follows. First evaluate the expression x[1](t) at the solution to pick out the $x_1(t)$ expression.

```
> eval( x[1](t), ans );
```

$$\frac{20}{11}\cos(\sqrt{11}\,t) + \frac{2}{11} + \frac{155}{121}\text{Heaviside}(t-1)$$

$$-\frac{100}{121}\text{Heaviside}(t-1)\cos(\sqrt{11}\,(t-1))$$

$$+\frac{5}{11}\text{Heaviside}(t-1)\,t^2 - \frac{10}{11}\text{Heaviside}(t-1)\,t$$

Then turn the expression into a function by using unapply.

```
> y[1] := unapply( %, t );
```

$$y_1 := t \to \frac{20}{11}\cos(\sqrt{11}\,t) + \frac{2}{11} + \frac{155}{121}\text{Heaviside}(t-1)$$

$$-\frac{100}{121}\text{Heaviside}(t-1)\cos(\sqrt{11}\,(t-1))$$

$$+\frac{5}{11}\text{Heaviside}(t-1)\,t^2 - \frac{10}{11}\text{Heaviside}(t-1)\,t$$

You can also do the two steps at once.

```
> y[2] := unapply( eval( x[2](t), ans ), t );
```

$$y_2 := t \to \frac{2}{11} - \frac{2}{11}\cos(\sqrt{11}\,t) + \frac{45}{121}\text{Heaviside}(t-1)$$

$$+\frac{5}{11}\text{Heaviside}(t-1)\,t^2 - \frac{10}{11}\text{Heaviside}(t-1)\,t$$

$$+\frac{10}{121}\text{Heaviside}(t-1)\cos(\sqrt{11}\,(t-1))$$

Now you can plot the two functions.

```
> plot( [ y[1](t), y[2](t) ], t=-3..6 );
```

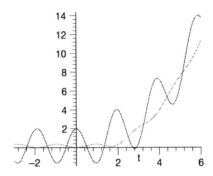

Instead of using dsolve(..., method=laplace), you may of course use the Laplace transform method by hand. The inttrans package defines the Laplace transform and its inverse (and many other integral transforms).

```
> with(inttrans);
```

[*addtable, fourier, fouriercos, fouriersin, hankel, hilbert,*

*invfourier, invhilbert, invlaplace, invmellin, laplace,*

*mellin, savetable*]

The Laplace transforms of the two differential equations eqn1 and eqn2 are

```
> laplace( eqn1, t, s );
```

$$\alpha\, m\, (s\, (s\, \text{laplace}(x_1(t),\ t,\ s)) - x_1(0)) - D(x_1)(0)) =$$
$$k\, (\text{laplace}(x_2(t),\ t,\ s) - \text{laplace}\, (x_1(t),\ t,\ s)) + \frac{e^{(-s)}}{s}$$

and

```
> laplace( eqn2, t, s );
```

$$m\, (s\, (s\, \text{laplace}(x_2(t),\ t,\ s)) - x_2(0)) - D(x_2)(0)) =$$
$$k\, (\text{laplace}(x_1(t),\ t,\ s) - \text{laplace}\, (x_2(t),\ t,\ s))$$

Evaluate the set consisting of the two transforms at the initial conditions.

```
> eval( {%, %%}, {ini} );
```

$$\{m\,s^2\,\text{laplace}(x_2(t),\,t,\,s) =$$

$$k\,(\text{laplace}(x_1(t),\,t,\,s) - \text{laplace}\,(x_2(t),\,t,\,s)),$$

$$\alpha\,m\,s\,(s\,\text{laplace}(x_1(t),\,t,\,s) - 2) =$$

$$k\,(\text{laplace}(x_2(t),\,t,\,s) - \text{laplace}\,(x_1(t),\,t,\,s)) + \frac{e^{(-s)}}{s}\}$$

You must now solve this set of algebraic equations for the Laplace transforms of the two functions $x_1(t)$ and $x_2(t)$.

```
> sol := solve( %, { laplace(x[1](t),t,s),
>       laplace(x[2](t),t,s) } );
```

$$sol := \{\text{laplace}(x_1(t),\,t,\,s) = \frac{(2\,\alpha\,m\,s^2\,e^s + 1)\,(m\,s^2 + k)}{e^s\,m\,s^3\,(\alpha\,m\,s^2 + \alpha\,k + k)},$$

$$\text{laplace}(x_2(t),\,t,\,s) = \frac{k\,(2\,\alpha\,m\,s^2\,e^s + 1)}{e^s\,m\,s^3\,(\alpha\,m\,s^2 + \alpha\,k + k)}\}$$

Maple has solved the algebraic problem; now you need to take the inverse
Laplace transform to get the functions $x_1(t)$ and $x_2(t)$ out.

```
> invlaplace( %, s, t );
```

$$\left\{ x_1(t) = \left( 2m\,\cos\left( \sqrt{\frac{k\,(\alpha+1)}{\alpha\,m}}\,t \right) + 2\,\frac{\alpha\,m}{\alpha+1} \right. \right.$$

$$-2\,\frac{\alpha\,m\,\cos\left( \sqrt{\frac{k\,(\alpha+1)}{\alpha\,m}}\,t \right)}{\alpha+1} + \frac{m\,\text{Heaviside}(t-1)}{k\,(\alpha+1)}$$

$$-\frac{m\,\text{Heaviside}(t-1)\cos\left( \sqrt{\frac{k\,(\alpha+1)}{\alpha\,m}}\,(t-1) \right)}{k\,(\alpha+1)}$$

$$-\frac{\text{Heaviside}(t-1)\,\alpha\,m}{k\,(\alpha+1)^2} + \frac{1}{2}\,\frac{\text{Heaviside}(t-1)\,t^2}{\alpha+1}$$

$$-\frac{\text{Heaviside}(t-1)\,t}{\alpha+1} + \frac{1}{2}\,\frac{\text{Heaviside}(t-1)}{\alpha+1}$$

$$
+ \frac{\text{Heaviside}(t-1)\,\alpha\,m \cos\left(\sqrt{\dfrac{k\,(\alpha+1)}{\alpha\,m}}\,(t-1)\right)}{k\,(\alpha+1)^2}\Bigg) \Big/ m,
$$

$$
x_2(t) = k\left(2\,\frac{\alpha\,m}{k\,(\alpha+1)} - 2\,\frac{\alpha\,m \cos\left(\sqrt{\dfrac{k\,(\alpha+1)}{\alpha\,m}}\,t\right)}{k\,(\alpha+1)}\right.
$$

$$
-\frac{\text{Heaviside}(t-1)\,\alpha\,m}{k^2\,(\alpha+1)^2} + \frac{1}{2}\,\frac{\text{Heaviside}(t-1)\,t^2}{k\,(\alpha+1)}
$$

$$
-\frac{\text{Heaviside}(t-1)\,t}{k\,(\alpha+1)} + \frac{1}{2}\,\frac{\text{Heaviside}(t-1)}{k\,(\alpha+1)}
$$

$$
\left. +\frac{\text{Heaviside}(t-1)\,\alpha\,m \cos\left(\sqrt{\dfrac{k\,(\alpha+1)}{\alpha\,m}}\,(t-1)\right)}{k^2\,(\alpha+1)^2}\right)\Big/ m\Bigg\}
$$

Evaluate at values for the constants.

```
> eval( %, {alpha=1/10, m=1, k=1} );
```

$$
\{x_1(t) = \frac{20}{11}\cos(\sqrt{11}\,t) + \frac{2}{11} + \frac{155}{121}\text{Heaviside}(t-1)
$$

$$
-\frac{100}{121}\text{Heaviside}(t-1)\cos(\sqrt{11}\,(t-1))
$$

$$
+\frac{5}{11}\text{Heaviside}(t-1)\,t^2 - \frac{10}{11}\text{Heaviside}(t-1)\,t, x_2(t) =
$$

$$
\frac{2}{11} - \frac{2}{11}\cos(\sqrt{11}\,t) + \frac{45}{121}\text{Heaviside}(t-1)
$$

$$
+\frac{5}{11}\text{Heaviside}(t-1)\,t^2 - \frac{10}{11}\text{Heaviside}(t-1)\,t
$$

$$
+\frac{10}{121}\text{Heaviside}(t-1)\cos(\sqrt{11}\,(t-1))\}
$$

As expected, you get the same solution as before.

**The type=series Option**  The series method for solving differential equations finds an approximate symbolic solution to the equations in the following manner. Maple finds a series approximation to the equations. It then solves the series approximation symbolically, using exact methods.

This technique is useful when Maple's standard algorithms fail, but you still want a symbolic solution rather than a purely numeric solution. The series method can often help with non-linear and high-order ODEs.

When using the series method, Maple assumes that a solution of the form

$$x^c \left( \sum_{i=0}^{\infty} a_i x^i \right)$$

exists, where $c$ is a rational number.

Consider the following differential equation.

```
> eq := 2*x*diff(y(x),x,x) + diff(y(x),x) + y(x) = 0;
```

$$eq := 2x \left( \frac{\partial^2}{\partial x^2} y(x) \right) + \left( \frac{\partial}{\partial x} y(x) \right) + y(x) = 0$$

Ask Maple to solve the equation.

```
> dsolve( {eq}, {y(x)}, type=series );
```

$$y(x) = \_C1 \sqrt{x}(1 - \frac{1}{3}x + \frac{1}{30}x^2 - \frac{1}{630}x^3 + \frac{1}{22680}x^4 -$$

$$\frac{1}{1247400}x^5 + O(x^6)) + \_C2$$

$$\left(1 - x + \frac{1}{6}x^2 - \frac{1}{90}x^3 + \frac{1}{2520}x^4 - \frac{1}{113400}x^5 + O(x^6)\right)$$

Use rhs to pick out the solution; then convert it to a polynomial.

```
> rhs(%);
```

$$\_C1 \sqrt{x}(1 - \frac{1}{3}x + \frac{1}{30}x^2 - \frac{1}{630}x^3 + \frac{1}{22680}x^4 - \frac{1}{1247400}$$

$$x^5 + O(x^6)) + \_C2$$

$$\left(1 - x + \frac{1}{6}x^2 - \frac{1}{90}x^3 + \frac{1}{2520}x^4 - \frac{1}{113400}x^5 + O(x^6)\right)$$

```
> poly := convert(%, polynom);
```

$$poly := \_C1 \sqrt{x}$$

$$\left(1 - \frac{1}{3}x + \frac{1}{30}x^2 - \frac{1}{630}x^3 + \frac{1}{22680}x^4 - \frac{1}{1247400}x^5\right)$$

$$+ \_C2 \left( 1 - x + \frac{1}{6}x^2 - \frac{1}{90}x^3 + \frac{1}{2520}x^4 - \frac{1}{113400}x^5 \right)$$

Now you can plot the solution for different values of the arbitrary constants _C1 and _C2.

```
> [ seq( _C1=i, i=0..5 ) ];
```

$$[\_C1 = 0, \ \_C1 = 1, \ \_C1 = 2, \ \_C1 = 3, \ \_C1 = 4, \ \_C1 = 5]$$

```
> map(subs, %, _C2=1, poly);
```

$$[1 - x + \frac{1}{6}x^2 - \frac{1}{90}x^3 + \frac{1}{2520}x^4 - \frac{1}{113400}x^5,$$

$$\%1 + 1 - x + \frac{1}{6}x^2 - \frac{1}{90}x^3 + \frac{1}{2520}x^4 - \frac{1}{113400}x^5,$$

$$2\,\%1 + 1 - x + \frac{1}{6}x^2 - \frac{1}{90}x^3 + \frac{1}{2520}x^4 - \frac{1}{113400}x^5,$$

$$3\,\%1 + 1 - x + \frac{1}{6}x^2 - \frac{1}{90}x^3 + \frac{1}{2520}x^4 - \frac{1}{113400}x^5,$$

$$4\,\%1 + 1 - x + \frac{1}{6}x^2 - \frac{1}{90}x^3 + \frac{1}{2520}x^4 - \frac{1}{113400}x^5,$$

$$5\,\%1 + 1 - x + \frac{1}{6}x^2 - \frac{1}{90}x^3 + \frac{1}{2520}x^4 - \frac{1}{113400}x^5]$$

$$\%1 :=$$

$$\sqrt{x} \left( 1 - \frac{1}{3}x + \frac{1}{30}x^2 - \frac{1}{630}x^3 + \frac{1}{22680}x^4 - \frac{1}{1247400}x^5 \right)$$

```
> plot( %, x=1..10 );
```

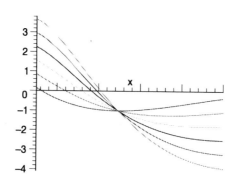

**The** type=numeric **Option**    Although the series methods for solving ODEs are well understood and adequate for finding accurate approximations of the dependent variable, they do exhibit some limitations. In order to obtain a result, the resultant series must converge; moreover, in the process of finding the solution Maple must find many derivatives, and that can be expensive in terms of time and memory. For these and other reasons, alternative numerical solvers have been developed.

Here is a differential equation and an initial condition.

```
> eq := x(t) * diff(x(t), t) = t^2;
```

$$eq := x(t) \left( \frac{\partial}{\partial t} x(t) \right) = t^2$$

```
> ini := x(1) = 2;
```

$$ini := x(1) = 2$$

The output from the dsolve command with the numeric option is a procedure which returns a list of equations when you call it.

```
> sol := dsolve( {eq, ini}, {x(t)}, type=numeric );
```

$$sol := \mathbf{proc}(rkf45\_x) \ \dots \ \mathbf{end}$$

The solution satisfies the initial condition.

```
> sol(1);
```

$$[t = 1, \ x(t) = 2.]$$

```
> sol(0);
```

$$[t = 0, \ x(t) = 1.82574187591285053]$$

Use the eval command to select a particular value from the list of equations.

```
> eval( x(t), sol(1) );
```

$$2.$$

You can also create an ordered pair.

```
> eval( [t, x(t)], sol(0) );
```

$$[0, \ 1.82574187591285053]$$

The plots package contains a command, odeplot, for plotting the result of dsolve( ..., type=numeric).

```
> with(plots):
```

```
> odeplot( sol, [t, x(t)], -1..2 );
```

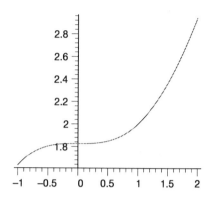

See ?plots,odeplot for the syntax of odeplot.
Here is a system of two ODEs.

```
> eq1 := diff(x(t),t) = y(t);
```

$$eq1 := \frac{\partial}{\partial t} x(t) = y(t)$$

```
> eq2 := diff(y(t),t) = x(t)+y(t);
```

$$eq2 := \frac{\partial}{\partial t} y(t) = x(t) + y(t)$$

```
> ini :=  x(0)=2, y(0)=1;
```

$$ini := x(0) = 2, \; y(0) = 1$$

This time the solution-procedure yields a list of three equations.

```
> sol1 := dsolve( {eq1, eq2, ini}, {x(t),y(t)},
>     type=numeric );
```

$$sol1 := \mathbf{proc}(rkf45\_x) \dots \mathbf{end}$$

```
> sol1(0);
```

$$[t = 0, \; x(t) = 2., \; y(t) = 1.]$$

```
> sol1(1);
```

$$[t = 1, \; x(t) = 5.58216868924484366,$$

$$y(t) = 7.82689113711079365]$$

The odeplot command can now plot y(t) against x(t),

```
> odeplot( sol1, [x(t), y(t)], -3..1, labels=["x","y"] );
```

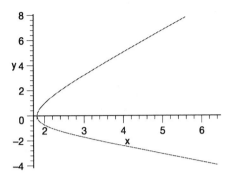

or x(t) and y(t) against t,

```
> odeplot( sol1, [t, x(t), y(t)], -3..1,
>     labels=["t","x","y"], axes=boxed );
```

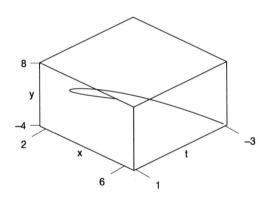

or any other combination.

Always use caution when using numeric methods. Consider the equation

```
> eq := diff(y(x), x) = 1 - 2*x*y(x);
```

$$eq := \frac{\partial}{\partial x} y(x) = 1 - 2 x y(x)$$

```
> ini := y(0) = 0;
```

$$ini := y(0) = 0$$

This differential equation happens to have an exact solution.

```
> exact := dsolve( {eq, ini}, {y(x)} );
```

$$exact := y(x) = -\frac{1}{2} I e^{(-x^2)} \sqrt{\pi} \, \text{erf}(I x)$$

```
> plot( rhs(exact), x=-5..5, title="Exact Solution" );
```

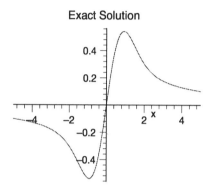

Exact Solution

However, if you use the type=numeric option, the graph is very different.

```
> approx := dsolve( {eq, ini}, {y(x)}, type=numeric );
```

$$approx := \text{proc}(rkf45\_x) \ldots \text{end}$$

```
> with(plots):
> odeplot( approx, [x,y(x)], -5..5, view=[-5..5, -0.6..0.6],
>      title="Numeric Solution" );
```

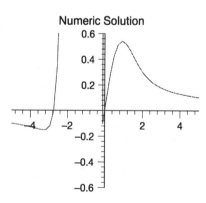

Numeric Solution

Errors can accumulate in floating-point calculations. Universal rules for preventing this effect do not exist, and so no software package can anticipate all conditions. The solution here is to use the startinit option to make dsolve (or rather the procedure which dsolve returns) begin at the initial value every time you calculate a point $(x, y(x))$.

```
> approx2 := dsolve( {eq, ini}, {y(x)},
>    type=numeric, startinit=true );
```

$$approx2 := \mathbf{proc}(rkf45\_x) \ldots \mathbf{end}$$

```
> odeplot( approx2, [x,y(x)], -5..5,
>    title="With startinit=true" );
```

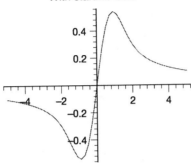

The cost is that approx2 is slower than approx.

You can specify which algorithm dsolve(..., type=numeric) should use when solving your differential equation; see ?dsolve,numeric.

## Example: Taylor Series

In its general form, a series method solution to an ODE requires the forming of a Taylor series about $t = 0$ for some function $f(t)$. Thus, you must be able to obtain and manipulate the higher order derivatives of your function, $f'(t)$, $f''(t)$, $f'''(t)$, and so on.

Once you have obtained the derivatives, you can substitute them into the Taylor series representation of $f(t)$.

```
> taylor(f(t), t);
```

$$f(0) + D(f)(0)\,t + \frac{1}{2}\,(D^{(2)})(f)(0)\,t^2 + \frac{1}{6}\,(D^{(3)})(f)(0)\,t^3 +$$

$$\frac{1}{24}\,(D^{(4)})(f)(0)\,t^4 + \frac{1}{120}\,(D^{(5)})(f)(0)\,t^5 + O(t^6)$$

As an example, consider Newton's Law of Cooling:

$$\frac{d\theta}{dt} = -\frac{1}{10}(\theta - 20), \qquad \theta(0) = 100.$$

Using the D operator, you can easily enter the above equation into Maple.

```
> eq := D(theta) = -1/10*(theta-20);
```

$$eq := D(\theta) = -\frac{1}{10}\theta + 2$$

```
> ini := theta(0)=100;
```

$$ini := \theta(0) = 100$$

The first step is to obtain the required number of higher derivatives. Coordinating this number with the order of the Taylor series that you are going to use is a sensible idea. If you use the default value of Order that Maple provides,

```
> Order;
```

$$6$$

then you must generate derivatives up to order

```
> dev_order := Order - 1;
```

$$dev\_order := 5$$

You can now use seq to generate a sequence of the higher order derivatives of theta(t).

```
> S := seq( (D@@(dev_order-n))(eq), n=1..dev_order );
```

$$S := (D^{(5)})(\theta) = -\frac{1}{10}(D^{(4)})(\theta), \ (D^{(4)})(\theta) = -\frac{1}{10}(D^{(3)})(\theta),$$

$$(D^{(3)})(\theta) = -\frac{1}{10}(D^{(2)})(\theta), \ (D^{(2)})(\theta) = -\frac{1}{10}D(\theta),$$

$$D(\theta) = -\frac{1}{10}\theta + 2$$

The fifth derivative is a function of the fourth derivative, the fourth a function of the third and so on. Therefore, if you make substitutions according to S, you may express all the derivatives as functions of theta. For example, the third element of S is the following.

```
> S[3];
```

$$(D^{(3)})(\theta) = -\frac{1}{10}(D^{(2)})(\theta)$$

Substituting according to S on the right-hand side, yields

```
> lhs(%) = subs( S, rhs(%) );
```

$$(D^{(3)})(\theta) = -\frac{1}{1000}\theta + \frac{1}{50}$$

To make this substitution on all the derivatives at once, use the map command.

```
> L := map( z -> lhs(z) = eval(rhs(z), {S}), [S] );
```

$$L := [(D^{(5)})(\theta) = \frac{1}{100}(D^{(3)})(\theta), \ (D^{(4)})(\theta) = \frac{1}{100}(D^{(2)})(\theta),$$

$$(D^{(3)})(\theta) = \frac{1}{100}D(\theta), \ (D^{(2)})(\theta) = \frac{1}{100}\theta - \frac{1}{5},$$

$$D(\theta) = -\frac{1}{10}\theta + 2]$$

You must evaluate the derivatives at $t = 0$.

```
> L(0);
```

$$[(D^{(5)})(\theta)(0) = \frac{1}{100}(D^{(3)})(\theta)(0),$$

$$(D^{(4)})(\theta)(0) = \frac{1}{100}(D^{(2)})(\theta)(0),$$

$$(D^{(3)})(\theta)(0) = \frac{1}{100}D(\theta)(0), \ (D^{(2)})(\theta)(0) = \frac{1}{100}\theta(0) - \frac{1}{5},$$

$$D(\theta)(0) = -\frac{1}{10}\theta(0) + 2]$$

Now generate the Taylor series.

```
> T := taylor(theta(t), t);
```

$$T := \theta(0) + D(\theta)(0)\,t + \frac{1}{2}(D^{(2)})(\theta)(0)\,t^2 + \frac{1}{6}(D^{(3)})(\theta)(0)$$

$$t^3 + \frac{1}{24}(D^{(4)})(\theta)(0)\,t^4 + \frac{1}{120}(D^{(5)})(\theta)(0)\,t^5 + O(t^6)$$

Substitute the derivatives into the series.

```
> subs( op(L(0)), T );
```

$$\theta(0) + \left(-\frac{1}{10}\theta(0) + 2\right)t + \left(\frac{1}{200}\theta(0) - \frac{1}{10}\right)t^2 +$$

$$\left(-\frac{1}{6000}\theta(0) + \frac{1}{300}\right)t^3 + \left(\frac{1}{240000}\theta(0) - \frac{1}{12000}\right)t^4 +$$

$$\left(-\frac{1}{12000000}\theta(0) + \frac{1}{600000}\right)t^5 + O(t^6)$$

Now, evaluate the series at the initial condition and convert it to a polynomial.

```
> eval( %, ini );
```

$$100 - 8t + \frac{2}{5}t^2 - \frac{1}{75}t^3 + \frac{1}{3000}t^4 - \frac{1}{150000}t^5 + O(t^6)$$

```
> p := convert(%, polynom);
```

$$p := 100 - 8t + \frac{2}{5}t^2 - \frac{1}{75}t^3 + \frac{1}{3000}t^4 - \frac{1}{150000}t^5$$

You can now plot the response.

```
> plot(p, t=0..30);
```

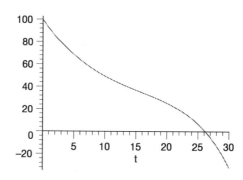

This particular example has the following analytic solution.

```
> dsolve( {eq(t), ini}, {theta(t)} );
```

$$\theta(t) = 20 + 80\,e^{(-1/10\,t)}$$

```
> q := rhs(%);
```

$$q := 20 + 80\,e^{(-1/10\,t)}$$

Thus, you can compare the series solution with the actual solution.

```
> plot( [p, q], t=0..30 );
```

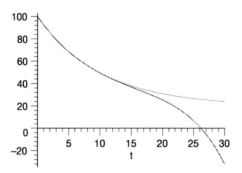

Instead of doing the Taylor series by hand, you may use the powsolve command from the powseries package.

```
> with(powseries);
```

[*compose, evalpow, inverse , multconst, multiply,*

    *negative, powadd, powcos, powcreate, powdiff, powexp,*

    *powint, powlog, powpoly, powsin, powsolve, powsqrt,*

    *quotient, reversion, subtract, tpsform*]

The powseries package requires that the differential equation is in diff-form rather than in D-form.

```
> eqn := convert( eq(t), diff );
```

$$eqn := \frac{\partial}{\partial t}\,\theta(t) = -\frac{1}{10}\,\theta(t) + 2$$

The powsolve command finds a formal power series solution.

```
> a := powsolve( {eqn, ini} );
```

$$a := \textbf{proc}(powparm) \ldots \textbf{end}$$

Now $a(n)$ is the coefficient of $x^n$ in the power series solution to eq. That is, you can derive the solution by typing

```
> Sum( a(n)*x^n, n=0..infinity );
```

$$\sum_{n=0}^{\infty} a(n)\, x^n$$

The constant term is

> a(0);

$$100$$

The general term is

> a(_k);

$$-\frac{1}{10}\frac{a(\_k-1)}{\_k}$$

The tpsform command returns the truncated power series form of a formal power series.

> tpsform(a, x);

$$100 - 8t + \frac{2}{5}t^2 - \frac{1}{75}t^3 + \frac{1}{3000}t^4 - \frac{1}{150000}t^5 + O(t^6)$$

## When You Cannot Find a Closed Form Solution

In some instances, you cannot express the solution to a linear ODE in closed form. In such cases, dsolve may return solutions containing the data structure DESol. DESol is a place holder representing the solution of a differential equation without explicitly computing it. Thus DESol is similar to RootOf, which represents the roots of an expression. This allows you to manipulate the resulting expression symbolically prior to attempting another approach.

> de := (x^7+x^3-3)*diff(y(x),x,x) + x^4*diff(y(x),x)
> + (23*x-17)*y(x);

$$de :=$$

$$(x^7 + x^3 - 3)\left(\frac{\partial^2}{\partial x^2}y(x)\right) + x^4\left(\frac{\partial}{\partial x}y(x)\right) + (23x - 17)y(x)$$

dsolve cannot find a closed form solution to de.

> dsolve( {de}, {y(x)} );

$$y(x) = \text{DESol}\left(\left\{(x^7 + x^3 - 3)\left(\frac{\partial^2}{\partial x^2}\_Y(x)\right) + x^4\left(\frac{\partial}{\partial x}\_Y(x)\right)\right.\right.$$

$$\left.\left. + (23x - 17)\_Y(x)\right\}, \{\_Y(x)\}\right)$$

You can now try another method on the DESol itself. For example, find a series approximation.

```
> series(rhs(%), x);
```

$$_Y(0) + D(_Y)(0) x - \frac{17}{6} \,_Y(0) x^2 +$$

$$\left( -\frac{17}{18} D(_Y)(0) + \frac{23}{18} \,_Y(0) \right) x^3 +$$

$$\left( \frac{289}{216} \,_Y(0) + \frac{23}{36} D(_Y)(0) \right) x^4 +$$

$$\left( \frac{289}{1080} D(_Y)(0) - \frac{833}{540} \,_Y(0) \right) x^5 + O(x^6)$$

diff and int can also operate on DESol.

## Plotting Ordinary Differential Equations

You cannot solve many differential equations analytically. In such cases, plotting the differential equation is advantageous.

```
> ode1 :=
> diff(y(t), t$2) + sin(t)^2*diff(y(t),t) + y(t) = cos(t)^2;
```

$$ode1 := \left( \frac{\partial^2}{\partial t^2} y(t) \right) + \sin(t)^2 \left( \frac{\partial}{\partial t} y(t) \right) + y(t) = \cos(t)^2$$

```
> ic1 := y(0) = 1, D(y)(0) = 0;
```

$$ic1 := y(0) = 1, D(y)(0) = 0$$

First, attempt to solve this ODE analytically using dsolve.

```
> dsolve({ode1, ic1}, {y(t)} );
```

dsolve returned nothing, indicating that it could not find a solution. Try Laplace methods.

```
> dsolve( {ode1, ic1}, {y(t)}, method=laplace );
```

Again, dsolve did not find a solution. Since dsolve was not successful, try the DEplot command found in the DEtools package.

```
> with(DEtools);
```

$$[DEnormal, \; DEplot, \; DEplot3d, \; DEplot\_polygon,$$

*DFactor, Dchangevar, GCRD, LCLM,*

*PDEchangecoords, PDEplot, RiemannPsols, abelsol,*

*adjoint, autonomous, bernoullisol, buildsol, buildsym,*

*canoni, chinisol, clairautsol, constcoeffsols, convertAlg,*

*convertsys, dalembertsol, de2diffop, dfieldplot,*

*diffop2de, eigenring, endomorphism_charpoly, equinv,*

*eta_k, eulersols, exactsol, expsols, exterior_power,*

*formal_sol, gen_exp, generate_ic, genhomosol,*

*hamilton_eqs, indicialeq, infgen, integrate_sols,*

*intfactor, kovacicsols, leftdivision, liesol, line_int,*

*linearsol, matrixDE, matrix_riccati, moser_reduce,*

*mult, newton_polygon, odeadvisor, odepde,*

*parametricsol, phaseportrait, poincare, polysols,*

*ratsols, reduceOrder, regular_parts, regularsp,*

*riccati_system, riccatisol, rightdivision, separablesol,*

*super_reduce, symgen, symmetric_power,*

*symmetric_product, symtest, transinv, translate,*

*untranslate, varparam, zoom]*

DEplot is a general ODE plotter which you can use with the following syntax.

```
DEplot( ode, dep-var, range, [ini-conds] )
```

Here *ode* is the differential equation you want to plot, *dep-var* is the dependent variable, *range* is the range of the independent variable, and *ini-conds* is a list of initial conditions.

Here is a plot of the function satisfying both the differential equation ode1 and the initial conditions ic1 above.

```
> DEplot( ode1, y(t), 0..20, [ [ ic1 ] ] );
```

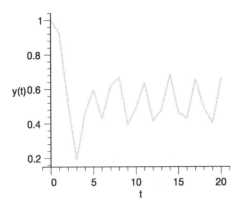

You can refine the plot by specifying a smaller stepsize.

```
> DEplot( ode1, y(t), 0..20, [ [ ic1 ] ], stepsize=0.2 );
```

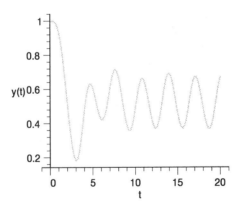

If you specify more than one list of initial conditions, DEplot plots a solution for each.

```
> ic2 := y(0)=0, D(y)(0)=1;
```

$$ic2 := y(0) = 0, \; D(y)(0) = 1$$

```
> DEplot( ode1, y(t), 0..20, [ [ic1], [ic2] ], stepsize=0.2 );
```

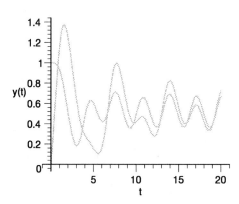

DEplot can also plot solutions to a set of differential equations.

```
> eq1 := diff(y(t),t) + y(t) + x(t) = 0;
```

$$eq1 := \left( \frac{\partial}{\partial t} y(t) \right) + y(t) + x(t) = 0$$

```
> eq2 := y(t) = diff(x(t), t);
```

$$eq2 := y(t) = \frac{\partial}{\partial t} x(t)$$

```
> ini1 := x(0)=0, y(0)=5;
```

$$ini1 := x(0) = 0, \; y(0) = 5$$

```
> ini2 := x(0)=0, y(0)=-5;
```

$$ini2 := x(0) = 0, \; y(0) = -5$$

The system {eq1, eq2} has two dependent variables, x(t) and y(t), so you must give a list of dependent variables.

```
> DEplot( {eq1, eq2}, [x(t), y(t)], -5..5,
> [ [ini1], [ini2] ] );
```

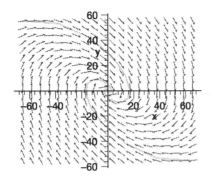

Note that DEplot also generates a direction field, as above, whenever it is meaningful to do so. See ?DEtools,DEplot for more details on how to plot ODEs.

DEplot3d is the three-dimensional version of DEplot. The basic syntax of DEplot3d is similar to that of DEplot. See ?DEtools,DEplot3d for details. Here is a three-dimensional plot of the system plotted in two dimensions above.

```
> DEplot3d( {eq1, eq2}, [x(t), y(t)], -5..5,
> [ [ini1], [ini2] ] );
```

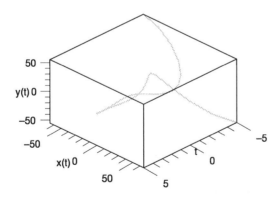

Here is an example of a plot of a system of three differential equations.

```
> eq1 := diff(x(t),t) = y(t)+z(t);
```

$$eq1 := \frac{\partial}{\partial t} x(t) = y(t) + z(t)$$

```
> eq2 := diff(y(t),t) = -x(t)-y(t);
```

$$eq2 := \frac{\partial}{\partial t} y(t) = -y(t) - x(t)$$

```
> eq3 := diff(z(t),t) = x(t)+y(t)-z(t);
```

$$eq3 := \frac{\partial}{\partial t} z(t) = x(t) + y(t) - z(t)$$

These are two lists of initial conditions.

```
> ini1 := [x(0)=1, y(0)=0, z(0)=2];
```

$$ini1 := [x(0) = 1, \, y(0) = 0, \, z(0) = 2]$$

```
> ini2 := [x(0)=0, y(0)=2, z(0)=-1];
```

$$ini2 := [x(0) = 0, \, y(0) = 2, \, z(0) = -1]$$

The DEplot3d command plots two solutions to the system of differential equations {eq1, eq2, eq3}, one solution for each list of initial values.

```
> DEplot3d( {eq1, eq2, eq3}, [x(t), y(t), z(t)], t=0..10,
>       [ini1, ini2], stepsize=0.1, orientation=[-171, 58] );
```

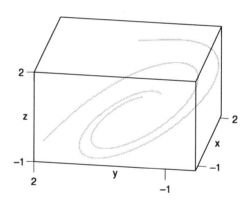

## Discontinuous Forcing Functions

In many practical instances the forcing function to a system is discontinuous. Maple provides a number of ways in which you can describe a system in terms of ODEs and include, in a meaningful way, descriptions of discontinuous forcing functions.

**The Heaviside Step Function** Using the Heaviside function allows you to model delayed and piecewise-defined forcing functions. You may use Heaviside with dsolve to find both symbolic and numeric solutions.

```
> eq := diff(y(t),t) = -y(t)*Heaviside(t-1);
```

$$eq := \frac{\partial}{\partial t} y(t) = -y(t) \, \text{Heaviside}(t - 1)$$

```
> ini := y(0) = 3;
```

$$ini := y(0) = 3$$

```
> dsolve({eq, ini}, {y(t)});
```

$$y(t) = 3\,e^{((-t+1)\text{Heaviside}(t-1))}$$

Turn the solution into a function which you can plot.

```
> rhs( % );
```

$$3\,e^{((-t+1)\text{Heaviside}(t-1))}$$

```
> f := unapply(%, t);
```

$$f := t \rightarrow 3\,e^{((-t+1)\text{Heaviside}(t-1))}$$

```
> plot(f, 0..4);
```

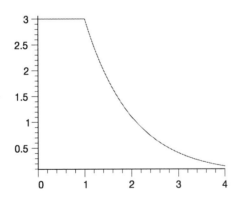

Solve the same equation numerically.

```
> sol1 := dsolve({eq, ini}, {y(t)}, type=numeric);
```

$$sol1 := \textbf{proc}(rkf45\_x) \ldots \textbf{end}$$

You can use the odeplot command from the plots package to plot the solution.

```
> with(plots):
```

```
> odeplot( sol1, [t, y(t)], 0..4 );
```

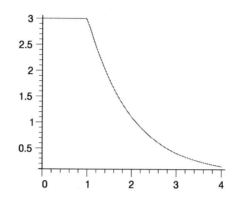

**The Dirac Delta Function**   You can use the Dirac delta function in a similar manner to produce impulsive forcing functions.

```
> eq := diff(y(t),t) = -y(t)*Dirac(t-1);
```

$$eq := \frac{\partial}{\partial t} y(t) = -y(t) \, \text{Dirac}(t-1)$$

```
> ini := y(0) = 3;
```

$$ini := y(0) = 3$$

```
> dsolve({eq, ini}, {y(t)});
```

$$y(t) = 3 \, e^{(-\text{Heaviside}(t-1))}$$

Turn the solution into a function which you can plot.

```
> f := unapply( rhs( % ), t );
```

$$f := t \to 3 \, e^{(-\text{Heaviside}(t-1))}$$

```
> plot( f, 0..4 );
```

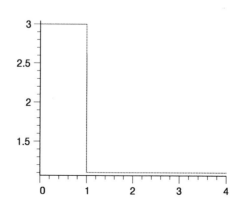

However, the numeric solution does not see the non-zero value of `Dirac(0)`.

```
> sol2 := dsolve({eq, ini}, {y(t)}, type=numeric);
```

$$sol2 := \mathbf{proc}(rkf45\_x) \ldots \mathbf{end}$$

Again, use `odeplot` from `plots` to plot the numeric solution.

```
> with(plots, odeplot);
```

$$[odeplot]$$

```
> odeplot( sol2, [t,y(t)], 0..4 );
```

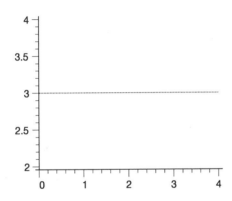

**Piecewise Functions**   The `piecewise` command allows you to construct complicated forcing functions by approximating sections of it with analytic functions, and then taking the approximations together to represent the whole function. First, look at the behavior of `piecewise`.

```
> f:=  x -> piecewise(1<=x and x<2, 1, 0);
```

$$f := x \to \text{piecewise}(1 \le x \text{ and } x < 2, 1, 0)$$

```
> f(x);
```

$$\begin{cases} 1, & \text{if}, 1 - x \le 0 \text{ and } x - 2 < 0; \\ 0, & \text{otherwise.} \end{cases}$$

Note that the order of the conditionals is important. The first conditional that returns `true` causes the function to return a value.

> plot(f, 0..3);

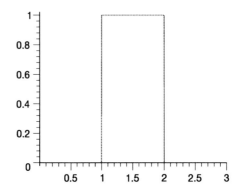

Thus, you can use this piecewise function as a forcing function.

> eq := diff(y(t),t) = 1-y(t)*f(t);

$$eq := \frac{\partial}{\partial t} y(t) = 1 - y(t) \left( \begin{cases} 1, & \text{if } 1 - t \leq 0 \text{ and } t - 2 < 0; \\ 0, & \text{otherwise.} \end{cases} \right)$$

> ini := y(0)=3;

$$ini := y(0) = 3$$

> sol3 := dsolve({eq, ini}, {y(t)}, type=numeric);

$$sol3 := \mathbf{proc}(rkf45\_x) \ldots \mathbf{end}$$

Again, use the odeplot command in the plots package to plot the result.

> with(plots, odeplot):
> odeplot( sol3, [t, y(t)], 0..4 );

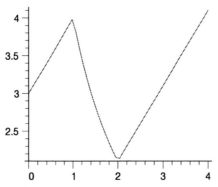

The DEtools package contains commands that can help you investigate, manipulate, plot, and solve differential equations. See ?DEtools for details.

## 6.3 Partial Differential Equations

Partial differential equations (PDEs) are in general very difficult to solve. Maple provides a number of commands for solving, manipulating, and plotting PDEs. Some of these commands are in the standard library, but most of them reside in the PDEtools package.

### The pdsolve Command

The pdsolve command can solve many partial differential equations. This is the basic syntax of the pdsolve command.

pdsolve( *pde, var* )

Here *pde* is the partial differential equation and *var* is the variable for which you want Maple to solve.

The following is the one-dimensional wave equation.

```
> wave := diff(u(x,t), t,t) - c^2 * diff(u(x,t), x,x);
```

$$wave := \left( \frac{\partial^2}{\partial t^2} u(x,\ t) \right) - c^2 \left( \frac{\partial^2}{\partial x^2} u(x,\ t) \right)$$

You want to solve for u(x,t). First load the PDEtools package.

```
> with(PDEtools):
> sol := pdsolve( wave, u(x,t) );
```

$$sol := u(x,\ t) = \_F2 \left( -\frac{\sqrt{c^2}\, x}{c^2} + t \right) + \_F1 \left( \frac{1}{2} x + \frac{1}{2} \frac{t\, c^2}{\sqrt{c^2}} \right)$$

Note the solution is in terms of two arbitrary functions, _F1 and _F2. To plot the solution you need a particular set of functions.

```
> f1 := xi -> exp(-xi^2);
```

$$f1 := \xi \rightarrow e^{(-\xi^2)}$$

```
> f2 := xi -> piecewise(-1/2<xi and xi<1/2, 1, 0);
```

$$f2 := \xi \rightarrow \text{piecewise} \left( \frac{-1}{2} < \xi \text{ and } \xi < \frac{1}{2},\ 1,\ 0 \right)$$

Substitute these functions into the solution.

```
> eval( sol, {_F1=f1, _F2=f2, c=1} );
```

$$u(x, t) = \left( \begin{cases} 1 & -\frac{1}{2} + x - t < 0 \text{ and } -x + t - \frac{1}{2} < 0 \\ 0 & \text{otherwise} \end{cases} \right.$$
$$+ e^{(-(1/2\,x+1/2\,t)^2)}$$

You can use the rhs command to pick out the solution.

```
> rhs(%);
```

$$\left( \begin{cases} 1 & -\frac{1}{2} + x - t < 0 \text{ and } -x + t - \frac{1}{2} < 0 \\ 0 & \text{otherwise} \end{cases} \right.$$
$$+ e^{(-(1/2\,x+1/2\,t)^2)}$$

unapply turns the expression into a function.

```
> f := unapply(%, x,t);
```

$$f := (x, t) \rightarrow$$
$$\text{piecewise} \left( -\frac{1}{2} + x - t < 0 \text{ and } -x + t - \frac{1}{2} < 0, 1, 0 \right)$$
$$+ e^{(-(1/2\,x+1/2\,t)^2)}$$

Now you can plot the solution.

```
> plot3d( f, -8..8, 0..5, grid=[40,40] );
```

## Changing the Dependent Variable in a PDE

Below is the one-dimensional heat equation.

```
> heat := diff(u(x,t),t) - k*diff(u(x,t), x,x) = 0;
```

$$heat := \left( \frac{\partial}{\partial t} u(x,\ t) \right) - k \left( \frac{\partial^2}{\partial x^2} u(x,\ t) \right) = 0$$

Try to find a solution of the form $X(x)T(t)$ to this equation. Use the aptly named HINT option of pdsolve to suggest a course of action.

```
> pdsolve( heat, u(x,t), HINT=X(x)*T(t));
```

$$(u(x,\ t) = X(x)\,T(t))\ \&\text{where}$$

$$\left[ \left\{ \frac{\partial}{\partial t} T(t) = k\,\_c_1\,T(t),\ \frac{\partial^2}{\partial x^2} X(x) = \_c_1\,X(x) \right\} \right]$$

The result here is correct, but difficult to read.

Alternatively, you can tell pdsolve to use separation of variables (as a product, '*') and then solve the resulting ODEs (using 'build' option).

```
> sol := pdsolve(heat, u(x,t), HINT='*', 'build');
```

$$sol := u(x,\ t) = \_C3\,e^{(k\,\_c_1\,t)}\,\_C1\,\sinh(\sqrt{\_c_1}\,x)$$

$$+\ \_C3\,e^{(k\,\_c_1\,t)}\,\_C2\,\cosh(\sqrt{\_c_1}\,x)$$

Evaluate the solution at specific values for the constants.

```
> S := eval( rhs(sol), {_C3=1, _C1=1, _C2=1, k=1, _c[1]=1} );
```

$$S := e^t\,\sinh(x) + e^t\,\cosh(x)$$

You can plot the solution.

```
> plot3d( S, x=-5..5, t=0..5 );
```

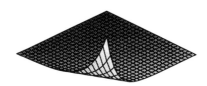

Checking the solution by evaluation with the original equation is a good idea.

```
> eval( heat, u(x,t)=rhs(sol) );
```

$$\_C3\, k\, \_c_1\, e^{(k\, \_c_1 t)}\, \_C1 \sinh(\sqrt{\_c_1}\, x)$$
$$+ \_C3\, k\, \_c_1\, e^{(k\, \_c_1 t)}\, \_C2 \cosh(\sqrt{\_c_1}\, x) - k($$
$$\_C3\, e^{(k\, \_c_1 t)}\, \_C1 \sinh(\sqrt{\_c_1}\, x)\, \_c_1$$
$$+ \_C3\, e^{(k\, \_c_1 t)}\, \_C2 \cosh(\sqrt{\_c_1}\, x)\, \_c_1) = 0$$

```
> simplify(%);
```

$$0 = 0$$

## Plotting Partial Differential Equations

The solutions to many PDEs can be plotted with the PDEplot command found in the PDEtools package.

```
> with(PDEtools):
```

You can use the PDEplot command with the following syntax.

> PDEplot( *pde*, *var*, *ini*, *s=range* )

Here *pde* is the PDE, *var* is the dependent variable, *ini* is a parametric curve in three-dimensional space with parameter *s*, and *range* is the range of *s*.

Consider this partial differential equation.

```
> pde := diff(u(x,y), x) + cos(2*x) * diff(u(x,y), y) = -sin(y);
```

$$pde := \left(\frac{\partial}{\partial x} u(x,\ y)\right) + \cos(2\,x) \left(\frac{\partial}{\partial y} u(x,\ y)\right) = -\sin(y)$$

Use the curve given by $z = 1 + y^2$ as an initial condition, that is $x = 0$, $y = s$, and $z = 1 + s^2$.

```
> ini := [0, s, 1+s^2];
```

$$ini := [0,\ s,\ 1 + s^2]$$

PDEplot draws the initial-condition curve (though it is difficult to see in this book) and the solution surface.

```
> PDEplot( pde, u(x,y), ini, s=-2..2 );
```

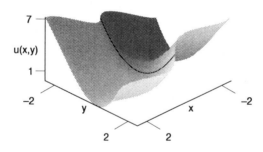

To draw the surface, Maple calculates these base characteristic curves. The initial-condition curve is easier to see here than in the above plot.

```
> PDEplot( pde, u(x,y), ini, s=-2..2, basechar=only );
```

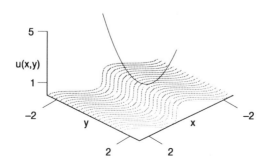

The basechar=true option tells PDEplot to draw both the character-istic curves and the surface, as well as the initial-condition curve which is always present.

```
> PDEplot( pde, u(x,y), ini, s=-2..2, basechar=true );
```

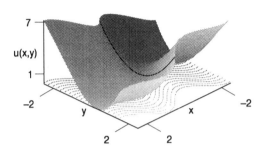

Many of the usual plot3d options are also available; see ?plot3d, options. The initcolor option sets the color of the initial value curve.

```
> PDEplot( pde, u(x,y), ini, s=-2..2,
>      basechar=true, initcolor=white,
>      style=patchcontour, contours=20,
>      orientation=[-43,45] );
```

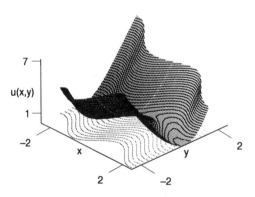

The difforms and liesymm packages both contain routines for sophis-ticated mathematical analysis of PDEs. See ?difforms and ?liesymm.

## 6.4 Conclusion

This chapter has demonstrated how useful Maple is when you want to investigate and solve problems using calculus. You have seen how Maple can illustrate concepts, such as the derivative and the Riemann integral; how Maple helps you through a careful analysis of the error term in a Taylor approximation; and Maple's powerful facilities for solving and manipulating ordinary and partial differential equations, numerically as well as symbolically.

# Input and Output

You can do much of your work directly within Maple's worksheets. You can perform calculations, plot functions, and document the results. However, at some point you may need to import data or export results to a file in order to interact with another person or piece of software. The data could be measurements from scientific experiments or numbers generated by other programs. Once you import the data into Maple, you can use Maple's plotting capabilities to visualize the results, and its algebraic capabilities to construct or investigate an associated mathematical model.

Maple provides a number of convenient ways to both import and export raw numerical data and graphics. It presents individual algebraic and numeric results in formats suitable for use in FORTRAN or C, or even mathematical typesetting systems such as LaTeX. You can even export the entire worksheet as a text file (for inclusion in electronic mail) or as a LaTeX document. You can cut and paste results, and export either single expressions or entire worksheets.

This chapter takes you through the most common aspects of exporting and importing information to and from files. It introduces how Maple interacts with the file system on your computer, and how Maple can begin interacting with other software.

## 7.1 Reading Files

The two most common reasons to read files are to obtain data and to retrieve Maple commands stored in a text file.

The first case is often concerned with data generated from an experiment. You can store numbers separated by white space and line breaks in a text file, then read them into Maple for study. You can most easily accomplish this operation using Maple's readdata command.

The second case is the reading of commands from a text file. Perhaps you receive a worksheet in text format, or write a Maple procedure using your favorite text editor and store it in a text file before running it. You can Cut and Paste commands into Maple or you can use the read command. *Reading Commands from a File* on page 262 discusses the latter option.

## Reading Columns of Numbers from a File

Maple is very good at manipulating data, but if you generate this data outside Maple, you must read it into Maple before you can manipulate it. Often such external data is in the form of columns of numbers in a text file. The file data.txt below is an example.

```
0  1  0
1  .540302  .841470
2 -.416146  .909297
3 -.989992  .141120
4 -.653643 -.756802
5  .283662 -.958924
6  .960170 -.279415
```

The readdata command reads such columns of numbers. Use readdata as follows.

```
readdata( "filename", n )
```

Here, *filename* is the name of the file that you want readdata to read, and *n* is the number of columns. If *n* is 1, then readdata returns a list of numbers. Otherwise, readdata returns a list of lists, each sublist corresponding to a line in the file.

The file data.txt has three columns.

```
> L := readdata( "data.txt", 3 );
```

$$L := [[0, 1., 0], [1., .540302, .841470],$$
$$[2., -.416146, .909297], [3., -.989992, .141120],$$
$$[4., -.653643, -.756802], [5., .283662, -.958924],$$
$$[6., .960170, -.279415]]$$

Now you can plot the third column against the first, for example. Use the map command to pick out the first and the third entries in each sublist.

```
> map( u -> [ u[1], u[3] ], L );
```

$$[[0, 0], [1., .841470], [2., .909297], [3., .141120],$$
$$[4., -.756802], [5., -.958924], [6., -.279415]]$$

The plot command can plot lists like this one directly.

```
> plot(%);
```

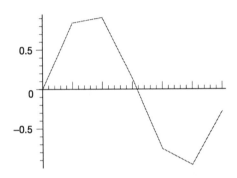

You can also do statistical analysis on your data using the commands in the stats package.

```
> with(stats);
```

$$[anova, \ describe, \ fit, \ importdata, \ random, \ statevalf,$$
$$statplots, \ transform]$$

Suppose you want to find the mean (average) of the second column. The describe subpackage of the stats package defines the mean command.

```
> with(describe);
```

$$[coefficientofvariation, \ count, \ countmissing, \ covariance,$$
$$decile, \ geometricmean, \ harmonicmean, \ kurtosis,$$
$$linearcorrelation, \ mean, \ meandeviation, \ median, \ mode,$$
$$moment, \ percentile, \ quadraticmean, \ quantile, \ quartile,$$
$$range, \ skewness, \ standarddeviation, \ sumdata, \ variance$$
$$]$$

You can use the map command as above to pick out the second column of numbers. Alternatively, you can use the fact that L[5,2] is the second number in the fifth sublist,

```
> L[5,2];
```

$$-.653643$$

So, the following is the list you need.

```
> L[ 1..nops(L), 2 ];
```

$$[1., .540302, -.416146, -.989992, -.653643, .283662,$$

$$.960170]$$

```
> mean(%) ;
```

$$.1034790000$$

If your imported data represents a matrix, you may want to convert a list of lists, such as L, to an array.

```
> A := convert(L, array);
```

$$A := \begin{bmatrix} 0 & 1. & 0 \\ 1. & .540302 & .841470 \\ 2. & -.416146 & .909297 \\ 3. & -.989992 & .141120 \\ 4. & -.653643 & -.756802 \\ 5. & .283662 & -.958924 \\ 6. & .960170 & -.279415 \end{bmatrix}$$

Now you can perform matrix calculations on your data. The evalm command evaluates a matrix expression.

```
> transpose(A) &* A;
```

$$transpose(A) \&^* A$$

```
> evalm(%);
```

$$\begin{bmatrix} 91. & 1.302792 & -6.414894 \\ 1.302792 & 3.874827634 & -.1090779271 \\ -6.414894 & -.1090779271 & 3.125164896 \end{bmatrix}$$

By default, readdata reads values as floating-point numbers, but it can also read values as integers. The file integers.txt consists of two columns of integers.

```
1 1
2 4
3 9
4 16
5 25
```

The `integer` keyword tells `readdata` to read values as integers instead of floating-point numbers.

```
> readdata( "integers.txt", integer, 2 );
```

$$[[1, 1], [2, 4], [3, 9], [4, 16], [5, 25]]$$

You can also mix integers and floating-point numbers. Suppose the first column consists of integers and the second column consists of floating-point numbers. If you specify the formats in a list, you do not have to specify the number of columns.

```
> readdata( "integers.txt", [integer, float] );
```

$$[[1, 1.], [2, 4.], [3, 9.], [4, 16.], [5, 25.]]$$

Do not use the `integer` keyword unless the numbers in the relevant column(s) actually are integers.


## Reading Commands from a File

Some Maple users find it convenient to write Maple programs in a text file with their favorite text editor, then import the file into Maple. You can paste the commands from the text file into your worksheet or you can use the `read` command.

When you read a file with the `read` command, Maple treats each line in the file as a command. Maple executes the commands and displays the results in your worksheet but it does *not*, by default, place the commands from the file in your worksheet. Use the `read` command with the following syntax.

```
read "filename";
```

Here is the file `ks.tst` of Maple commands.

```
S := n -> sum( binomial(n, beta)
    * ( (2*beta)!/2^beta - beta!*beta ), beta=1..n );
S( 19 );
```

When you read the file in, Maple displays the results but not the commands.

```
> read "ks.tst";
```

$$S := n \rightarrow \sum_{\beta=1}^{n} \text{binomial}(n,\ \beta) \left( \frac{(2\,\beta)!}{2^{\beta}} - \beta!\,\beta \right)$$

$$1024937361666644598071114328769317982974$$

If you set the `interface` variable `echo` to 2, Maple inserts the commands from the file into your worksheet.

```
> interface( echo=2 );
> read "ks.tst";

> S := n -> sum( binomial(n, beta)
>     * ( (2*beta)!/2^beta - beta!*beta ), beta=1..n );
```

$$S := n \rightarrow \sum_{\beta=1}^{n} \text{binomial}(n,\ \beta) \left( \frac{(2\,\beta)!}{2^{\beta}} - \beta!\,\beta \right)$$

```
> S( 19 );
```

$$1024937361666644598071114328769317982974$$

The `read` command can also read files in Maple's internal format; see *Saving Expressions in Maple's Internal Format* on page 265.

## 7.2 Writing Data to a File

After using Maple to perform a calculation, you may want to save the result in a file. You can then process the result later, either with Maple or with another program.

### Writing Columns of Numerical Data to a File

If the result of a Maple calculation is a long list or a large array of numbers, you may want to write these numbers to a file in a structured manner. The `writedata` command writes columns of numerical data, allowing you to import the numbers into some other program. You can use the `writedata` command with the following syntax.

```
writedata( "filename", data )
```

Here, *filename* is the string containing the name of the file where `writedata` puts the data, and *data* is a list, vector, list of lists, or matrix. If *filename* is the special name (not string) `terminal`, then `writedata` writes the data onto your screen. Note that `writedata` overwrites *filename* if it exists.

However, you may use the following syntax to append data to an existing file.

> writedata[APPEND]( "*filename*", *data* )

If the data is a vector or a list of numbers, then writedata writes one number per line.

```
> L := [ 3, 3.1415, -65, 0 ];
```

$$L := [3, 3.1415, -65, 0]$$

```
> writedata( terminal, L );

3
3.1415
-65
0
```

If the data is a matrix or a list of lists of numbers, then writedata writes columns of data and tab characters separate the columns.

```
> A := [ [1,2,3], [-1.45, 0, 3/2] ];
```

$$A := \left[ [1, 2, 3], \left[ -1.45, 0, \frac{3}{2} \right] \right]$$

```
> writedata( terminal, A );
1               2               3
-1.45           0               1.5
```

writedata expects that the data is numeric. You must evaluate constants, such as $\pi$ and $e^9$, to floating-point numbers before calling writedata.

```
> L := [ Pi, exp(9) ];
```

$$L := [\pi, e^9]$$

```
> Lf := evalf(L);
```

$$Lf := [3.141592654, 8103.083928]$$

```
> writedata( terminal, Lf );

3.141593
8103.083928
```

By default, writedata writes floating-point numbers. The keyword integer tells writedata to truncate the numbers to integers.

```
> L := [ 3.7, 3.1, -2.2, -1.9 ];
```

$$L := [3.7, 3.1, -2.2, -1.9]$$

```
> writedata( terminal, L, integer );
```

```
3
3
-2
-1
```

Thus, writedata(..., integer) truncates in a similar manner to the trunc command.

```
> map( trunc, L );
```

$$[3, \ 3, \ -2, \ -1]$$

If you want to convert floating-point numbers to integers in some other way, using round for example, you must use both the conversion command and the integer keyword.

```
> writedata( terminal, map(round, L), integer );
```

```
4
3
-2
-2
```

You can also mix integers and floating-point numbers. Below, the writedata command writes the first and third columns as floating-point numbers and the second column as integers.

```
> writedata( terminal, A, [float, integer, float] );
```

```
1               2        3
-1.45           0        1.5
```

You can extend writedata so that it will write more complicated data, such as complex numbers or symbolic expressions. See ?writedata for more information.

## Saving Expressions in Maple's Internal Format

If you construct a complicated expression or procedure, you may want to save it for future use in Maple. If you save the expression or procedure in Maple's internal format, then Maple can retrieve it efficiently. You can accomplish this by using the save command to write the expression to a file whose name ends with the characters ".m". Use the save command with the following syntax.

```
save nameseq, "filename.m";
```

Here *nameseq* is a sequence of names; you can only save named objects. The save command saves the objects in *filename*.m. The .m indicates that save will write the file using Maple's internal format.

Here are a few expressions.

```
> qbinomial := (n,k) -> product(1-q^i, i=n-k+1..n) /
>                       product(1-q^i, i=1..k );
```

$$qbinomial := (n, k) \rightarrow \frac{\displaystyle\prod_{i=n-k+1}^{n} (1 - q^i)}{\displaystyle\prod_{i=1}^{k} (1 - q^i)}$$

```
> expr := qbinomial(10, 4);
```

$$expr := \frac{(1 - q^7)\,(1 - q^8)\,(1 - q^9)\,(1 - q^{10})}{(1 - q)\,(1 - q^2)\,(1 - q^3)\,(1 - q^4)}$$

```
> nexpr := normal( expr );
```

$$nexpr := (q^6 + q^5 + q^4 + q^3 + q^2 + q + 1)\,(q^4 + 1)\,(q^6 + q^3 + 1)$$
$$(q^8 + q^6 + q^4 + q^2 + 1)$$

You can now save these expressions to the file qbinom.m.

```
> save qbinomial, expr, nexpr, "qbinom.m";
```

The restart command clears the three expressions from memory. Thus expr evaluates to its own name below.

```
> restart:
> expr;
```

$$expr$$

Use the read command to retrieve the expressions that you saved in qbinom.m.

```
> read "qbinom.m";
```

Now expr has its value again.

```
> expr;
```

$$\frac{(1 - q^7)\,(1 - q^8)\,(1 - q^9)\,(1 - q^{10})}{(1 - q)\,(1 - q^2)\,(1 - q^3)\,(1 - q^4)}$$

See *Reading Commands from a File* on page 262 for more on the read command.

## Converting to LATEX Format

TEX is a program for typesetting mathematics, and LATEX is a macro package for TEX. The `latex` command converts Maple expressions to LATEX format. Thus, you can use Maple to solve a problem, then convert the result to LATEX code which you can then include in a LATEX document. Use the `latex` command in the following manner.

```
latex( expr, "filename" )
```

The `latex` command writes the LATEX code corresponding to the Maple expression *expr* to the file *filename*. If *filename* exists, `latex` overwrites it. You may prefer to omit the *filename*; in that case `latex` prints the LATEX code on the screen and you can cut and paste it into your LATEX document.

```
> latex( a/b );
```

```
{\frac {a}{b}}
```

```
> latex( Limit( int(f(x), x=-n..n), n=infinity ) );
```

```
\lim _{n\rightarrow \infty }\int _{-n}^{n}\!f(
x){dx}
```

The `latex` command does not supply the commands necessary to tell LATEX to process the code in math-mode, nor does it attempt any line breaking or alignment.

*Export as LATEX* on page 269 describes how you can save a whole worksheet in LATEX format.

## 7.3 Exporting Whole Worksheets

You can, of course, save your worksheets through the Save or Save As... options of the File menu. However, you can also export a worksheet in four other formats: plain text, Maple text, LATEX, and HTML, through Export As submenu, which is also under the File menu. This allows you to process a worksheet outside Maple.

## Plain Text

You can save a worksheet as plain text by choosing Export As from the File menu and Plain Text from the ensuing submenu. In this case, Maple precedes input with a greater-than sign and a space (> ). Maple uses character-based typesetting for special symbols like integral signs and exponents, but you cannot export graphics as text. The following is a portion of a Maple worksheet exported in plain text format.

```
An Indefinite Integral
by Jane Maplefan
Calculation
Look at the integral Int(x^2*sin(x-a),x);. Notice that
its integrand, x^2*sin(x-a);, depends on the parameter
a;.
Give the integral a name so that you can refer to it
later.
> expr := Int(x^2 * sin(x-a), x);
```

$$ expr := \int x^2 \, \sin(x-a) \, dx $$

```
The value of the integral is an anti-derivative of the
integrand.
> answer := value( % );
```

## Maple Text

*Maple text* is specially marked text that retains the worksheets distinction between text, Maple input, and Maple output. Thus, you can export a worksheet as Maple text, send the text file by electronic mail, and the recipient can import the Maple text into a Maple session and regenerate most of the structure of your original worksheet. When reading or pasting Maple text, Maple treats each line that begins with a Maple prompt and a space (> ) as Maple input, each line that begins with a hash mark and a space (# ) as text, and ignores all other lines.

You can export a whole worksheet as Maple text by choosing Export As from the File menu and Maple Text from the ensuing submenu. The following is a potion of a Maple worksheet exported as Maple text.

```
# An Indefinite Integral
# by Jane Maplefan
# Calculation
# Look at the integral Int(x^2*sin(x-a),x);. Notice that
# its integrand, x^2*sin(x-a);, depends on the parameter
# a;.
# Give the integral a name so that you can refer to it
# later.
> expr := Int(x^2 * sin(x-a), x);
```

$$ \int $$
$$ |\quad 2 $$

```
expr :=  |  x  sin(x - a) dx
         |
        /
```

```
# The value of the integral is an anti-derivative of the
# integrand.
> answer := value( % );
```

To open a worksheet in Maple text format as the one above, choose Open from the File menu. In the dialog box that appears, choose Maple Text from the drop list of file types. Double-click on the desired file and choose *Maple Text* again from the ensuing dialog.

You can also copy and paste Maple text through the Edit menu. If you copy a part of your worksheet as Maple text and paste it into another application, then the pasted text appears as Maple text. Similarly, if you paste Maple text into your worksheet using Paste Maple Text from the Edit menu, then Maple retains the structure of the Maple text. In contrast, if you use ordinary paste, Maple does not retain any structure: if you paste into an input region, Maple interprets all the pasted text as input, and if you paste into a text region, Maple interprets all the pasted text as text.

## LaTeX

You can export a Maple worksheet in LaTeX format by choosing Export as from the File menu and LaTeX from the ensuing submenu. The .tex file that Maple generates is ready for processing by LaTeX. All distributions of Maple include the necessary style files.

If your worksheet contains embedded graphics, then Maple generates PostScript files corresponding to the graphics, as well as LaTeX code for including these PostScript files in your LaTeX document.

The following is a portion of a Maple worksheet exported as LaTeX.

```
%% Created by Maple VR5 (IBM INTEL NT)
%% Source Worksheet: tut1.mws
%% Generated: Sun Sep 07 10:10:45 1997
\documentclass{article}
\usepackage{maple2e}
\DefineParaStyle{Author}
\DefineParaStyle{Heading 1}
\DefineParaStyle{Maple Output}
\DefineParaStyle{Maple Plot}
\DefineParaStyle{Title}
\DefineCharStyle{2D Comment}
\DefineCharStyle{2D Math}
\DefineCharStyle{2D Output}
```

```
\DefineCharStyle{Hyperlink}
\begin{document}
\begin{maplegroup}
\begin{Title}
An Indefinite Integral
\end{Title}
\begin{Author}
by Jane Maplefan
\end{Author}
\end{maplegroup}
\section{Calculation}
Look at the integral
\mapleinline{inert}{2d}{Int(x^2*sin(x-a),x);}{%
$\int x^{2}\,\mathrm{sin}(x - a)\,dx$%
}. Notice that its integrand,
\mapleinline{inert}{2d}{x^2*sin(x-a);}{%
$x^{2}\,\mathrm{sin}(x - a)$%
}, depends on the parameter
\mapleinline{inert}{2d}{a;}{%
$a$%
}.
```

The LaTeX style files assume that you are printing the .tex file using the dvips printer driver. You can change this default by specifying an option to the \usepackage LaTeX command in the preamble of your .tex file.

*Printing Graphics* on page 271 describes how to save graphics directly. You can include such graphics files in your LaTeX document using the \mapleplot LaTeX command.

## HTML

You can export a Maple worksheet in HTML (HyperText Markup Language) format by choosing Export as from the File menu and HTML from the ensuing submenu. The .html file that Maple generates can be loaded into any HTML browser (for example, Netscape, Explorer, etc.) that supports frames.

Maple generates .gif files to represent equations and graphics in your worksheet.

The following is a Maple worksheet exported as HTML. Notice that other HTML documents (including a table of contents), which were created when you exported the original worksheet, are called from within it.

```
<html>
<head>
```

```
<title>tut1.htm</title>
<!-- Created by Maple V R5, IBM INTEL NT -->
</head>
<frameset cols="25%,*">
  <frame src="tut1TOC.htm" name="TableOfContents">
  <frame src="tut11.htm" name="Content">
<noframes>
Sorry, this document requires that your browser support
frames. <a href="tut11.htm" target="Content">This link</a>
will take you to a non-frames presentation of the
document.
</noframes>
</frameset>
</html>
```

The following is a portion of the tut11.htm file mentioned in the above file.

```
<b><font color=#000000 size=5>Calculation</font></b>
</p>
<p align=left>
<font color=#000000>Look at the integral </font>
<img src="tut11.gif" width=120 height=60 alt="[Maple Math]"
align=middle>
<font color=#000000>. Notice that its integrand, </font>
<img src="tut12.gif" width=89 height=50 alt="[Maple Math]"
align=middle>
<font color=#000000>, depends on the parameter </font>
<img src="tut13.gif" width=13 height=32 alt="[Maple Math]"
align=middle>
<font color=#000000>.</font>
</p>
<p align=left>
<font color=#000000>Give the integral a name so that you
can refer to it later.</font>
</p>
```

## 7.4 Printing Graphics

On most platforms, Maple by default displays graphics directly in the worksheet—as *in-line plots*. You can use the plotsetup command to change this behavior. The following command tells Maple to display graphics in separate windows on your terminal.

```
> plotsetup(window);
```

With your plot in a separate window, you can print it through the File menu as you would print any other worksheet.

The plotsetup command has the following general syntax.

```
plotsetup( DeviceType, plotoutput="filename",
plotoption="options" )
```

Here, *DeviceType* is the graphics device that Maple should use, *filename* is the name of the output file, and *options* is a string of options that the graphics driver recognizes.

The following command tells Maple to send graphics in PostScript format to the file myplot.ps.

```
> plotsetup( postscript, plotoutput="myplot.ps" );
```

The plot that the plot command below generates does not appear on the screen but, instead, goes to the file myplot.ps.

```
> plot( sin(x^2), x=-4..4 );
```

Maple can also generate graphics in a form suited to an HP Laserjet printer. Maple sends the graph that the plot3d command generates below to the file myplot.hp.

```
> plotsetup( hpgl, plotoutput="myplot.hp",
>                plotoptions=laserjet );
> plot3d( tan(x*sin(y)), x=-Pi/3..Pi/3, y=-Pi..Pi);
```

If you want to print more than one plot, you must change the plotoutput option between each plot; otherwise, the new plot overwrites the old one.

```
> plotsetup( plotoutput="myplot2.hp" );
> plot( exp@sin, 0..10 );
```

When you are done exporting graphics, you must tell Maple to send future graphics to your worksheet again.

```
> plotsetup( inline );
```

See ?plot,device for a description of the plotting devices that Maple knows.

## 7.5 Conclusion

In this chapter, you have seen a number of Maple's elementary input and output facilities: how to print graphics, how to save and retrieve individual

Maple expressions, how to read and write numerical data, and how to export a Maple worksheet as a LaTeX or HTML document.

In addition, Maple has many low-level input and output commands, such as `fprintf`, `fscanf`, `writeline`, `readbytes`, `fopen`, and `fclose`. See the help pages for more details.

The help pages are Maple's interactive reference manual. They are always at your fingertips when you are using Maple. Like a traditional reference manual, use them by studying the index, or by searching through them. In particular, the complete text search facility provides a method of searching for information, superior to a traditional index. In addition, hyperlinks make it easy for you to check other related topics.

This book aims to supply you with a good base of knowledge from which to explore Maple. In this role, it focuses on the interactive use of Maple. Of course, Maple is a complete language, and provides complete facilities for programming. In fact, the majority of Maple's commands are coded in the Maple language, as this high-level, mathematically oriented language is far superior to traditional computer languages for such tasks. The *Maple V Programming Guide* introduces you to programming in Maple.

# Index